REGIONAL INSECURITY: DHS GRANTS TO THE NATIONAL CAPITAL AREA

HEARING

BEFORE THE

COMMITTEE ON GOVERNMENT REFORM HOUSE OF REPRESENTATIVES

ONE HUNDRED NINTH CONGRESS

SECOND SESSION

JUNE 15, 2006

Serial No. 109–193

Printed for the use of the Committee on Government Reform

AF770825

Available via the World Wide Web: http://www.gpoaccess.gov/congress/index.html
http://www.house.gov/reform

U.S. GOVERNMENT PRINTING OFFICE

30–591 PDF WASHINGTON : 2006

For sale by the Superintendent of Documents, U.S. Government Printing Office
Internet: bookstore.gpo.gov Phone: toll free (866) 512–1800; DC area (202) 512–1800
Fax: (202) 512–2250 Mail: Stop SSOP, Washington, DC 20402–0001

CONTENTS

REGIONAL INSECURITY: DHS GRANTS TO THE NATIONAL CAPITAL AREA

THURSDAY, JUNE 15, 2006

HOUSE OF REPRESENTATIVES,
COMMITTEE ON GOVERNMENT REFORM,
Washington, DC.

The committee met, pursuant to notice, at 10 a.m., in room 2154, Rayburn House Office Building, Hon. Tom Davis (chairman of the committee) presiding.

Present: Representatives Tom Davis, Mica, Gutknecht, Platts, Foxx, Schmidt, Fossella, Wolf, Waxman, Maloney, Cummings, Kucinich, Clay, Watson, Van Hollen, Ruppersberger, Higgins, Norton, and Moran.

Staff present: David Marin, staff director; Keith Ausbrook, chief counsel; John Hunter, counsel; Rob White, communications director; Andrea LeBlanc, deputy director of communications; Shalley Kim and Wimberly Fair, professional staff members; Teresa Austin, chief clerk; Sarah D'Orsie, deputy clerk; Leneal Scott, computer systems manager; Kirstin Amerling; minority general counsel; Karen Lightfoot, minority communications director/senior policy advisor; Michael McCarthy and Kim Trinca, minority counsels; Richard Butcher, minority professional staff member; Earley Green, minority chief clerk; and Jean Gosa, minority assistant clerk.

Chairman TOM DAVIS. The meeting will come to order.

I would ask unanimous consent that Mr. Moran, Mr. Wynn, Mr. Cardin, Mr. Hoyer, Mr. Fossella and Mr. Wolf be able to participate in today's hearing. Without objection, it is so ordered.

The committee will come to order. On May 31st, the Department of Homeland Security announced fiscal year 2006 grant allocations for States and eligible high-risk urban areas. For the first time, grant awards reflected the use of relative risk and effectiveness criteria in an effort to better target Federal investments and make measurable progress toward the National Preparedness Goal. The new elaborate evaluation process used by DHS produced some unexpected, seemingly punitive results, particularly for the National Capital Region. We convene this morning to shed some needed light on how DHS assessed risks and needs in a region that has already been attacked and remains an attractive target for terrorists.

Without question, the Nation's Capital bears a disproportionate burden in terms of public safety challenges and Homeland Security costs. Comprised of 12 local jurisdictions, two States and the District of Columbia, this region must be prepared to protect critically important facilities and monuments of high operational and symbolic value to the entire Nation: the White House, the Pentagon, the Congress, the Supreme Court, to name just a few. The tragic events of September 11, 2001, and the anthrax attacks that same year unfortunately confirmed our unwelcome status as a prime target.

Yet one discretionary DHS grant program allocated on the basis of risk, the Urban Areas Security Initiative, yielded a reduced award to the National Capital Region this year: $47 million. Last year, the region received $77 million. True, Congress appropriated 14 percent less for the program this year, but the 40 percent reduction suffered by the NCR is clearly the product of something more than tight budgets.

According to DHS, this region stands in the top 25 percent of urban areas at risk in terms of both critical assets and geographic vulnerabilities. But the effectiveness of the proposed grant expenditures was ranked by DHS reviewers in the bottom 25 percent of all similar investment strategies. Leaders in this region and in other high-risk jurisdictions like New York and Los Angeles are asking, what happened? Why did some grant justifications score so poorly under the DHS system? And how much did secret reviews of unquestionably subjective factors, like relevance, innovative necessary and feasibility, undermine efforts to address real security needs?

In attempting to implement a risk-based grant allocation system and improve State and local response capabilities, DHS appears to have built a dangerous house of mirrors for the unwary, an overly elaborate system of marginally relevant evaluation criteria that equates the risk of terrorism with the risk of filing a bad grant application. The system seems to have taken little or no account of the most obvious indicator of risk imaginable, that the Nation's Capital has already been attacked.

Last week, the chairman of the 9/11 Commission, appointed by President Bush, former New Jersey Governor Thomas keen, said the grant awards to D.C. and New York appeared to "defy common sense." The process so far has also defied clear explanation, as DHS officials have offered different accounts of what factors produced the surprising allocations and what applicants might have done to improve their chances.

So today we seek to bring greater transparency to an important Homeland Security program. The effort to apply sound risk analysis and risk management standards to Homeland Security grants is commendable, and overdue. Scarce resources need to be focused on development of tangible and sustainable preparedness and response capabilities. But regional readiness to meet the threat of terrorism is not enhanced by wide, unpredictable and disruptive funding swings. At-risk cities and regions need to know their grant applications are being fully and fairly evaluated. The rules of the game should be clear. Now, grant applicants can only guess at the outcome of "black box" procedures fueled by classified threat information and secret peer reviews.

The Department has proposed a sustained and detailed debate on the difficult process of assessing risk, evaluating need and judging the effectiveness of local plans to build capabilities. We take them up on that pledge. Officials from the National Capital Region are also here to give their perspectives on their application and its evaluation. We look forward to a productive discussion of how DHS and this region plan to work together to address the National Capital area's unique security needs.

Let me just remind everybody, for the inaugural costs last year, 2 years ago, the money came out of the Department of Homeland Security grant money to the region. This was a burden shouldered by the city as well. Nothing was taken into account there, no reimbursements except to take it out of Homeland Security moneys at that point.

I would now recognize the distinguished ranking member, Mr. Waxman, for his opening statement.

[The prepared statement of Chairman Tom Davis follows:]

STATEMENT
CHARIMAN TOM DAVIS
COMMITTEE ON GOVERNMENT REFORM
"REGIONAL INSECURITY: DEPARTMENT OF HOMELAND SECURITY GRANTS TO THE NATIONAL CAPITAL AREA"

JUNE 15, 2006

On May 31, the Department of Homeland Security announced fiscal year 2006 grant allocations for states and eligible high-risk urban areas. For the first time, grant awards reflected the use of relative risk and effectiveness criteria in an effort to better target federal investments and make measurable progress toward the National Preparedness Goal. The new, elaborate evaluation process used by DHS produced some unexpected, seemingly punitive results, particularly for the National Capital Region (NCR). We convene this morning to shed some needed light on how DHS assessed risks and needs in a region that has already been attacked, and remains an attractive target for terrorists.

Without question the Nation's Capital bears a disproportionate burden in terms of public safety challenges and homeland security costs. Comprised of twelve local jurisdictions, two states and the District of Columbia, this region must be prepared to protect critically important facilities and monuments of high operational and symbolic value to the entire nation: the White House, the Pentagon, the Congress, the Supreme Court, to name just a few. The tragic events of September 11, 2001 and the anthrax attacks that same year unfortunately confirmed our unwelcome status as a prime target.

Yet one discretionary DHS grant program allocated on the basis of risk, the Urban Areas Security Initiative, yielded a reduced award to the National Capital Region this year: 47 million dollars. Last year the region received 77 million dollars. True, Congress appropriated fourteen percent less for the program this year. But the forty percent reduction suffered by the NCR is clearly the product of something more than tight budgets.

According to DHS, this region stands in the top 25 percent of urban areas at risk in terms of both critical assets and geographic vulnerabilities. But the effectiveness of the proposed grant expenditures was ranked by DHS reviewers in the *bottom* 25 percent of all similar investment strategies. Leaders in this region, and in other high risk jurisdiction like New York and Los Angeles, are asking: What happened? Why did some grant justifications score so poorly under the DHS system? And how much did secret reviews of unquestionably subjective factors like "relevance," "innovativeness," and "feasibility" undermine efforts to address real security needs?

In attempting to implement a risk-based grant allocation system and improve state and local response capabilities, DHS appears to have built a dangerous house of mirrors for the unwary, an overly elaborate system of marginally relevant evaluation criteria that equates the risk of terrorism with the risk of filing a "bad" grant application. The system seems to have taken little or no account of the most obvious indicator of risk imaginable - that the Nation's Capital has already been attacked.

Last week, the Chairman of the 9/11 Commission, former New Jersey governor Thomas Kean said the grant awards to DC and New York appeared to "defy common sense." The process so far has also defied clear explanation, as DHS officials have offered different accounts of what factors produced the surprising allocations and what applicants might have done to improve their chances.

So today we seek to bring greater transparency to an important homeland security program. The effort to apply sound risk analysis and risk management standards to homeland security grants is commendable, and overdue. Scarce resources need to be focused on development of tangible and sustainable preparedness and response capabilities. But regional readiness to meet the threat of terrorism is not enhanced by wide, unpredictable and disruptive funding swings. At-risk cities and regions need to know their grant applications are being fully and fairly evaluated. The rules of the game should be clear. Now, grant applicants can only guess at the outcome of "black box" procedures fueled by classified threat information and secret peer reviews.

The Department has promised a sustained and detailed debate on the difficult process of assessing risk, evaluating need and judging the effectiveness of local plans to build capabilities. We take them up on that pledge. Officials from the National Capital Region are also here to give their perspectives on their application and its evaluation. We look forward to a productive discussion of how DHS and this region plan to work together to address the Capital area's unique security needs.

Mr. WAXMAN. Thank you, Mr. Chairman, for holding this hearing, examining the drastic cuts in Homeland Security funding for the National Capital Region. The reduction in funding is sudden and unexpected, and it presents a serious challenge to the State and local officials work to protect this area. It also calls into question the priorities of this administration.

The Department of Homeland Security says that one reason for the cuts to D.C. is that the level of the risk for the region is lower relative to other cities according to classified computer programs and matrixes that the DHS uses. Well, this is difficult to understand. The Nation's Capital has enormous symbolic importance and is the center of our government and has already been the sight of a deadly attack. It is hard to see how it could be downgraded like this, and it is hard to have confidence in the secret system that the Department is using to rank risk when it comes to conclusions like this.

Ever since September 11th, I have been asking the administration to provide Congress with a comprehensive threat and risk assessment that we could use to inform budget decisions. To date, we have received a lot of rhetoric about how important it is to make sound risk-based decisions, but we have never seen the proof.

Local officials also suffer from the lack of transparency at the Department of Homeland Security. According to DHS, another reason for the cuts to D.C. was that the region's proposal scored poorly for effectiveness. But officials from D.C., Maryland and Virginia will testify today that they are baffled about why their proposals received low marks from DHS. Ironically, one of the criteria for effectiveness that DHS claims to examine is sustainability. How could local officials build sustainable programs when DHS funding for them varies wildly from year to year?

To protect our Nation, we need to think outside the narrow box the administration wants to put us in. The Department is proposing to increase funding for Los Angeles, which is absolutely right. The answer to the underfunding of D.C. is not to take funds from other high priorities. Instead, we need to look at our national priorities.

Last week, the Senate tried to pass a complete repeal of the estate tax. That boondoggle for the wealthiest families in the United States would have cost over $1 trillion. Yet we are being told today that we can't afford $30 million more to protect the Nation's Capital from terrorist attacks.

Earlier this year, we learned that auditors had challenged over $250 million in charges by Halliburton, yet the administration went ahead and paid them anyway. If we had saved that money, we would have had more than enough to fully fund Homeland Security grants to all of our urban areas.

Just this week, we learned that FEMA paid more than $1 billion in fraudulent claims in the aftermath of Katrina because the administration failed to have even basic controls in place. If we had spent these funds responsibly, we could have paid for the protections this region desperately needs.

Ultimately, the problem our Nation faces isn't the lack of funds; it is a lack of accountability and a failure of government to be responsible stewards of taxpayers' dollars. I am glad we are having this hearing on how to protect the National Capital Region, and I hope we can get answers about why funding was cut so dramatically and how the threats to this area can be addressed. I also hope the administration and Congress will reassess their spending priorities so we can truly make this Nation safer.

[The prepared statement of Hon. Henry A. Waxman follows:]

Statement of Rep. Henry A. Waxman, Ranking Minority Member
Committee on Government Reform
Hearing on "Regional Insecurity: DHS Grants to the National Capital Area"

June 15, 2006

Thank you, Mr. Chairman, for holding this hearing examining the drastic cuts in homeland security funding for the National Capital region.

The reduction in funding is sudden and unexpected, and it presents a serious challenge to the state and local officials working to protect this area. It also calls into question the priorities of this Administration.

The Department of Homeland Security says that one reason for the cuts to D.C. is that the risk level for the region is now lower relative to other cities, according to the classified computer programs and matrixes that DHS uses.

This is difficult to understand. The nation's capital has enormous symbolic importance and is the center of our government. And it's already been the site of a deadly attack. It's hard to see how it could it could be downgraded like this. And it's hard to have confidence in the secret system that the Department is using to rank risks when it comes to conclusions like this.

Ever since 9/11, I have been asking the Administration to provide Congress with a comprehensive threat and risk assessment that we could use to inform budget decisions. To date, we have received a lot of rhetoric about how important it is to make sound risk-based decisions. But we have never seen the proof.

Local officials also suffer from the lack of transparency at DHS. According to DHS, another reason for the cuts to D.C. was that the region's proposals scored poorly for effectiveness. But officials from D.C., Maryland, and Virginia will testify today that they are baffled about why their proposals received low marks from DHS. Ironically, one of the criteria for effectiveness that DHS claims to examine is sustainability. How can local officials build sustainable programs when DHS funding for them varies wildly from year to year?

To protect our nation, we need to think outside the narrow box the Administration wants to put us in. The Department is proposing to increase funding for Los Angeles, which is absolutely right. The answer to the underfunding of D.C. is not to rob funds from other high priorities.

Instead, we need to look at our national priorities. Last week, the Senate tried to pass a complete repeal of the estate tax. That boondoggle for the wealthiest families in the United States would have cost over a trillion dollars. Yet we are being told today that we can't afford $30 million more to protect the nation's capital from terrorist attacks.

Earlier this year, we learned that auditors had challenged over $250 million in charges by Halliburton. Yet the Administration went ahead and paid them anyway. If we had saved that money, we would have had more than enough to fully fund homeland security grants to all our urban areas.

And just this week, we learned that FEMA paid more than a billion dollars in fraudulent claims in the aftermath of Katrina because the Administration failed to have even basic controls in place. If we had just spent these funds responsibly, we could have paid for the protections this region desperately needs.

Ultimately, the problem our nation faces isn't a lack of funds. It's a lack of accountability and a failure of government to be responsible stewards of taxpayer dollars.

I'm glad we are having this hearing on how to protect the National Capital Region, and I hope we get answers about why funding was cut so dramatically and how the threats to this area can be addressed. But I also hope that the Administration and Congress will reassess their spending priorities so that we can truly make the nation safer.

Chairman TOM DAVIS. Thank you very much.

Ms. Norton.

Ms. NORTON. Thank you very much, Mr. Chairman.

I appreciate that you called this hearing. I believe, as I will indicate in my own statement, that the hearing you called today is a public service not only for this region but for our country because of the responsibilities of security officials in localities in this region.

Actually, this is the first of two hearings by committees on which I serve seeking answers to the same question: How could a risk-based security formula lead to unusually large reductions in Homeland Security funds for al Qaeda's favorite targets? Among them the National Capital Region, the sight of the September 11th attack on the Pentagon and site of the Capitol, located on this campus, where analysts believe the plane that was brought down in Pennsylvania was headed?

The Homeland Security Committee has asked the mayors of New York City and the District of Columbia to testify on this same subject next Wednesday. I have asked for a face-to-face meeting between regional officials and Department of Homeland Security headquarters staff, not only to get a greater understanding of the basis for the DHS decisions but particularly to understand how the region is expected to meet its unique dual obligations as officials charged with preventing and responding both for the Federal Government and local communities.

When the DHS grants were announced, the initial despair and bewilderment of residents and officials in this region quickly escalated to anger and outrage. The decision that was as astonishing as it was counter-intuitive was made with little explanation. We seek that explanation today.

Regional residents and Federal employees alike have admirably learned to live with higher risks than their fellow Americans, in part because of the special effort they see from regional and Federal security officials and the funding they assumed would come annually. Among these residents are 200,000 Federal employees from whom are drawn scores of thousands directly charged with the principal responsibility for designing and planning the National Homeland Security effort.

This region ranks high as a target, not only because of its almost 4 million residents, but unlike other targeted areas, the entire Federal presence, including the Nation's priceless iconic monuments and the Capitol, the Supreme Court and the White House and the Pentagon, all extremely inviting targets, are concentrated in a few square miles in the District of Columbia, Maryland and Virginia. Although individual sites are protected by Federal police and security guards, prevention and response for the region where the Federal presence is located is up to local officials and first responders, not the Federal Government.

When the Pentagon was attacked on September 11th, this Federal installation was completely dependent upon the largest fire department in the region, the D.C. Fire Department and other regional fire services and a helicopter and burn unit at Washington Hospital Center, among other local services.

The Nation learned that local officials here had a unique double responsibility for their own residents as well as for the Nation's most valuable and priceless Federal assets. And if I may say so, the Congress learned that the District of Columbia had almost total responsibility after the anthrax attack in the Congress of the United States, not the Federal Government, no agency of the Federal Government. They were sent to the old D.C. General Hospital for the District of Columbia to figure it out when we had an unprecedented attack right here on the campus of the Congress of the United States.

It is this twin responsibility that accounts for this region's unique obligations. It is this twin responsibility that most seriously raises questions about the judgment of the Department of Homeland Security in severely cutting the region's security funding. Even with the funding the region has received, I have spent more time than I can say worrying that there has not been enough time or funding to staunch, major and obvious vulnerabilities that mainly affect Federal employees who work here, not my own constituents. Working with appropriators, not DHS, I have gotten some but not all of the funding for ER One, for Washington Hospital Center, the only place where visitors to the Capitol and others in the region could be taken and isolated if there is a biological, chemical or other major such attack in Washington.

The tunnels, bridges and Metro rail and bus system that are responsible for bringing Federal workers to and from Federal jobs in the District every day present major untouched and difficult security issues, such as preventing and responding to attacks and accidents involving chemical, biological and other hazardous substances and fires from explosives in tunnels and the tunnels leading to this city.

After September 11th, a lack of interoperability here should be unthinkable, but that is a work in progress that requires additional funding. Perhaps most telling, given the many Federal buildings here where controversial decisions are made every day, is the lack of necessary equipment even for area bomb squads, not one of which has the equipment necessary to meet FEMA's top standard.

We will hear of other vulnerabilities today. We will want to know whether they were considered when DHS funding decisions were made, and we will want to know what precisely is now expected from local security officials charged with major prevention and recovery operations in the National Capital Region.

We will want to know not only because of those we represent; we will want to know because of the weighty, unique burden local security officials carry for the region where the Nation's most highly targeted assets are located.

Thank you, Mr. Chairman.

[The prepared statement of Hon. Eleanor Holmes Norton follows:]

ELEANOR HOLMES NORTON
DISTRICT OF COLUMBIA

COMMITTEE ON
HOMELAND SECURITY
SUBCOMMITTEES:
EMERGENCY PREPAREDNESS, SCIENCE
AND TECHNOLOGY
PREVENTION OF NUCLEAR AND
BIOLOGICAL ATTACK

Congress of the United States
House of Representatives
Washington, D.C. 20515

COMMITTEE ON
TRANSPORTATION AND
INFRASTRUCTURE
SUBCOMMITTEES:
RANKING MEMBER, ECONOMIC DEVELOPMENT,
PUBLIC BUILDINGS, AND
EMERGENCY MANAGEMENT
AVIATION
WATER RESOURCES AND
ENVIRONMENT

COMMITTEE ON
GOVERNMENT REFORM
SUBCOMMITTEES:
FEDERAL WORKFORCE AND
AGENCY ORGANIZATION
CRIMINAL JUSTICE, DRUG POLICY AND
HUMAN RESOURCES

Statement of Congresswoman Eleanor Holmes Norton at
Government Reform Full Committee Hearing
"Regional Insecurity-DHS Grants to the National Capital Area"

Thursday, June 15, 2006

This is the first of two hearings by committees on which I serve seeking answers to the same question: How could a risk-based security formula lead to unusually large reductions in homeland security funds for Al Qaeda's favorite targets, among them the National Capital Region, site of the 9/11 attack on the Pentagon and site of the Capitol located on this campus where analysts believe the plane that was brought down in Pennsylvania was headed? The Homeland Security Committee has asked the mayors of New York City and the District of Columbia to testify on this same subject next Wednesday. I have asked for a face-to-face meeting between regional officials and Department of Homeland Security headquarters staff, not only to get a greater understanding of the basis for the DHS decisions, but particularly to understand how the region is expected to meet its unique, dual obligations as officials charged with preventing and responding both for the federal government and local communities.

When the DHS grants were announced, the initial despair and bewilderment of residents and officials in this region quickly escalated to anger and outrage. The decision that was as astonishing as it was and counterintuitive was made with little explanation. We seek that explanation today. Regional residents and federal employees alike have admirably learned to live with higher risks than their fellow Americans, in part because of the special effort they see from regional and federal security officials and the funding that seemed to come annually. Among these residents are 200,000 federal employees from whom are drawn scores of thousands of federal employees directly charged with the principle responsibility for designing and planning the national homeland security effort.

This region ranks high as a target not only because of its almost four million residents. But unlike other targeted areas, the entire federal presence, including the nation's priceless, iconic monuments and the Capitol, the Supreme Court, the White House and the Pentagon-- are concentrated in a few square miles in the District of Columbia, Maryland, and Virginia. Although individual sites are protected by federal police and security guards, prevention and response for the region where the federal

529 14TH STREET, N.W., SUITE 900
WASHINGTON, D.C. 20045–1928
(202) 783-5065
(202) 783-5211 (FAX)

2136 RAYBURN HOUSE OFFICE BUILDING
WASHINGTON, D.C. 20515-5101
(202) 225-8050
(202) 225-3002 (FAX)
(202) 225-7829 (TDD)
www.norton.house.gov

2041 MARTIN L. KING AVENUE, S.E.
SUITE 300
WASHINGTON, D.C. 20020-5734
(202) 678-8900
(202) 678-8844 (FAX)

presence is located is up to local officials and first responders, not the federal government. When the Pentagon was attacked on 9/11, this federal installation was completely dependent upon the largest fire department in the region, the D.C. Fire Department, other regional fire services and the helicopter and burn unit at Washington Hospital Center, among other local emergency services. The nation learned that local officials here have a unique double responsibility for their own residents as well as for the nation's most valuable and priceless federal assets. It is this twin responsibility that accounts for this region's unique obligations. It is this twin responsibility that most seriously raises questions about the judgment of the Department of Homeland Security in severely cutting the region's security funding.

Even with the funding the region has received, I have spent more time than I can say worrying that there has not been enough time or funding to staunch major and obvious vulnerabilities that mainly affect federal employees, not my own constituents. Working with appropriators, not DHS, I have gotten some but not all of the funding for ER One for Washington Hospital Center, where visitors to the Capitol and others in the region could be taken and isolated if there is a biological, chemical or other major attack in Washington. The tunnels, bridges, and Metro rail and bus system that are responsible for bringing federal workers to and from federal jobs in the District every day present major, untouched and difficult security issues, such as preventing and responding to attacks and accidents involving chemical, biological and other hazardous substances and fires from explosives in WMATA tunnels and the tunnels leading into this city. After 9/11, a lack of interoperability here should be unthinkable, but that is a work in progress that requires additional funding. Perhaps most telling, given the many federal buildings here where controversial decisions are made, is the lack of necessary equipment even for area bomb squads, not one of which has the equipment necessary to meet FEMA's top standard.

We will hear of other vulnerabilities today, we will want to know whether they were considered when DHS funding decisions were made, and what precisely is now expected from local security officials charged with major prevention and recovery obligations in the National Capital Region. We will want to know not only because of those we represent. We will want to know because of the weighty, unique burden local security officials carry for the region where the nation's most highly targeted assets are located.

Chairman TOM DAVIS. Thank you very much.

Mrs. Maloney, any opening statement?

Mrs. MALONEY. Thank you very much, Chairman Davis and Ranking Member Waxman, for holding this very important hearing. My only wish is that this hearing would not only focus on D.C.'s Homeland Security but also New York's Homeland Security. Both D.C. and New York City, their funding was cut by 40 percent, yet by all lists, they remain target No. 1 in this Nation. The al Qaeda ranks them as target No. 1. Bin Laden has attacked New York City twice and has been caught on tape saying he intends to do it again. Likewise, the Capital Region has been attacked and suffered many other reported attempts to attack the Capital Region.

Since the announcement of the Homeland Security funding 2 weeks ago, I have heard a number of reasons why New York and Washington suffered such a tremendous cut. If you listen to the variety of explanations and excuses coming from Secretary Chertoff, your head will start spinning. He often contradicts himself.

We have heard that New York City and D.C. had their funding cut because overall funding for the Urban Area Security Initiative has been cut, the high threat funding formula. On June 7th, Secretary Chertoff was quoted in the paper as saying, "Congress gave us about $600 million less than our grant programs, including $125 million less for the Urban Areas Security Initiative." While it is true high-threat funding was cut by 15 percent, New York and D.C.'s funding was cut by 40 percent each while other communities saw increases of over 70 percent.

Also the initial formula had seven high-threat cities. That has now been expanded to 40 high-threat cities. Yet the 40 high-threat cities are not on the al Qaeda list, according to intelligence that has been released to Congress and others.

In fact, the cut to New York City was $83 million alone, or more than two-thirds of the total nationwide cut. When we have questioned the jump in funding for other communities, Tracy Henke, the DHS official in charge of the process, told me, "You have to understand that there is risk throughout the Nation."

I understand that risk exists outside of New York and D.C., but I would have preferred to have followed the recommendations of Secretary Chertoff when he stated in January that high threat grants are "not party favors to be distributed as widely as possible."

Then there is the much talked about listing of national monuments and icons or lack thereof. Yet the administration still thinks that New York has no icons, and certainly D.C. has more than any city in the country. I would like them to take a tour of New York. The Secretary's excuse for this is that they listed the Empire State Building, for example, as a large office building and the Brooklyn Bridge as a bridge. He contrasts that with Mount Rushmore, a monument with much fewer visitors.

To that I ask, can't our icons be listed as both? The Empire State Building is a large office building with thousands of workers, and it is also a very famous site that attracts hundreds of thousands of visitors and attention from terrorists. Al Qaeda was not videotaping the Brooklyn Bridge because it is a bridge. They were doing so because it is a bridge and a national icon on which thousands of people traverse each day.

It is this kind of common sense that seems to have completely escaped DHS as it tried to come up with a formula that is as complicated as possible.

The next reason the Department has been giving us as to why New York and Washington, DC, received a cut is due to our application. We both rank at the top for risk but at the bottom for our application.

Believe me, if there is one thing D.C. and New York know how to do, it is paperwork. They know how to fill out an application. They probably file more papers on more issues in more areas than any area in the country. So I don't know about you, Mr. Chairman, but I am going to ask the city of New York to maybe hire Steven King next time to write our application. Maybe then the Department will understand the full threat to New York City. And D.C., as my colleague mentioned, has suffered other attacks, the anthrax attacks, as has New York City.

So what started as a promising commitment from Secretary Chertoff at the beginning of this year that Homeland Security funding would, in fact, be based on risk has turned into one of the greatest displays of incompetency, and really, we cannot afford to be incompetent in the protection of our people and in our Homeland Security grants.

I look forward to the testimony. Thank you.

[The prepared statement of Hon. Carolyn B. Maloney follows:]

Congresswoman Carolyn Maloney
Committee on Government Reform
"Regional Insecurity: DHS Grants to the National Capital Area"
Room 2154, Rayburn House Office Building
June 15, 2006
10:00 a.m.

I would like to thank Chairman Davis
for holding today's hearing.

My only wish is that this hearing
would not only focus on the cuts
to homeland security funding
in our National Capitol Region,
but across the country.

I certainly think we all could learn
a thing or two about why New York City
has also seen a 40% reduction in
high-threat funding.

Since the announcement of
Homeland Security Funding two weeks ago,
I have heard a number of reasons
why New York City and Washington, D.C.
has suffered a cut.

If you listen to the variety of explanations and excuses from Secretary Chertoff and his staff your head will start spinning.

They often contradict themselves.

We have heard New York and D.C. had their funding cut because overall funding for the Urban Area Security Initiative (UASI) has been cut.

On June 7th, Secretary Chertoff told the New York Times:

[quote] "Congress gave us about $600 million less for our grant programs, including $125 million less for the urban areas initiative." [unquote]

While it is true high-treat funding

was cut by about 15%,

New York and D.C.'s funding was cut by 40% each, while other communities saw increases of 70% or more.

In fact the cut to New York City was $83 million alone, or more than 2/3 of the total nationwide cut.

When we have questioned the jump in funding for other communities, Tracy Henke, the DHS official in charge of the process tells us [quote] "We have to understand that there is risk throughout the nation". [unquote]

I certainly understand that risk exist outside of New York City, but I would have preferred to have followed the recommendation of Secretary Chertoff when he stated back in January that high-threat grants are “not party favors to be distributed as widely as possible”.

And then there’s the much talked-about listing of “national monuments and icons” – or lack thereof.
If the administration still thinks we have no icons after all this, I’d love to take them on a tour of New York – I’ll tell them they can meet me in Times Square.

The Secretary’s excuse for this one is that they listed the Empire State Building, for example, as “large office building” and the Brooklyn Bridge as “bridge.”

He contrasts them with Mt. Rushmore, a monument with much fewer vistors.

To that I ask, can’t our icons be listed as both?

The Empire State Building is a large office building with thousands of workers, and it's also a famous site that attracts many more visitors and attention from terrorists.

Al Qaeda wasn't videotaping the Brooklyn Bridge because it's just a bridge, they were doing so because it's a bridge AND is a national icon.

It's this kind of common sense that seems to have completely escaped DHS as it tried to come up with as complicated a formula as possible.

One other reason the Department has been giving as to why New York City and Washington, D.C. received a cut is due to our application.

We both rank at the top for risk,
but at the bottom for our application.

I don't know about you Mr. Chairman, but I am going to ask the City of New York to hire Steven King next year to write our application.

Maybe then the Department will understand the full threat New York City faces.

So what started as a promising commitment from Secretary Chertoff at the beginning of this year that Homeland Funding would finally be based on risk, has turned into one of the greatest displays of incompetence I have ever seen and the bottom line is that none of us feel any safer.

Again, I thank the Chairman for holding today's hearing and I look forward to the testimony of our witnesses.

Chairman TOM DAVIS. Thank you very much.

Mr. Moran.

Mr. MORAN. Thank you, Mr. Chairman. Thanks for holding this hearing and for staying on top of this through the Government Reform Committee.

It does boggle the mind that we would be cutting expenditures for Homeland Security by $690 million, that we would be taking $122 million out of the Urban Area Security Initiative, and then we would be paying for that $120 million cut by taking it all out of Washington, DC, and New York City.

I mean, this is, as I say, it boggles the mind. I hear the excuses, and that is what we are going to talk about today. But, you know, you put it in the larger context: What are we doing cutting Homeland Security by $690 million? We are going to debate today and tomorrow 10 hours on the topic of Iraq. Well, Iraq, and it has been said and will be said hundreds of times today, I trust, had nothing to do with September 11th. Saddam did not attack this country. He was not harboring terrorists. And until we went in and invaded the country and occupied it, he was no threat whatsoever to the United States of America. And yet it is relevant because we are spending $300 million a day in Iraq, almost $9 billion a month in Iraq. And how are we paying for it? We take $690 million out of this entire Nation's Homeland Security funds.

We had an amendment in the Appropriations Committee, the Homeland Security Appropriations bill. Mr. Obey offered it. What it did was to completely pay for all of the 9/11 Commission's recommendations that the President endorsed and the Congress endorsed. It would have paid for all of it: all of the public transit monitoring; all of the port security; instead of checking 5 percent of the containers, we would have checked 100 percent of the containers for radiological, chemical or biological threats, rail security, on and on. It would have paid for all of it.

You know how we were going to offset it? The people who earned more, more than $1 million a year in this country, get $114,000 annual tax cuts in addition to all the other tax cuts they get. This is just the new tax cuts we passed. Instead of getting a $114,000 tax cut a year, they get $101,000; a $13,000 reduction to people making more in tax cuts than $1 million a year. That would have paid for all of the 9/11 Commission's Homeland Security recommendations. Now, it was defeated on a party line vote, no surprise, but my point is, where is this Nation's priorities?

Now, you are not going to be able to answer that, but we have some people from DHS here. They might relay that message to some of their bosses, that there was some questioning about the Nation's priorities.

Now, within DHS, you tell us that you didn't like the grant applications. I think Mrs. Maloney makes a pretty good point that not to identify the Brooklyn Bridge as anything more than a bridge was questionable, not to identify the Statue of Liberty as a national icon is questionable, and not to recognize the priority that Washington, DC, must have when we have millions of people here, 20 million visitors a year, the seat of the Nation's Capital, of the entire free world, and you make a 40 percent reduction in the funding?

Now, maybe you don't like the application. But I worked in the executive branch for a few years. We thought—this was during the Nixon administration, and one of our roles was to sit down with the local government and work with them, because the objective is not to grade papers; it is to do what is in the best interests of the American people, in this case, to figure out how best to secure our Nation's Capital. There was $190 million additional needs identified, and the Department of Homeland Security decided to address 6.5 percent of those needs, from what I see, down from almost 10 percent.

You recognize that D.C. is at the top of the terrorism target list. To put it at the bottom of funding, these are things that need some answers, and I don't think it is unreasonable for us to be asking these questions.

Again, what we are going to be doing is shooting the messenger. It is the people that sent you here who make these top decisions in terms of allocating our Nation's resources who are the ones who ought to be on the hot seat. But, nevertheless, this hearing needs to be held. More importantly, we need to get some answers.

Mr. Chairman, again, I appreciate your holding the hearing.

[The prepared statement of Hon. James P. Moran follows:]

Congressman James P. Moran
Statement for the Record
June 15, 2006

Mr. Chairman, I want to thank you for holding this important hearing today. The citizens of this region, which includes my own congressional district, are all very concerned about this year's significant cut in the National Capital Region's Homeland Security grant allocation. I greatly appreciate the opportunity to participate in this hearing.

The National Capital Region (NCR) is one of the most at-risk regions in our country. Washington, D.C. and its surrounding suburbs are the home of approximately 4.2 million people and destination of 20 million tourist visitors annually. As the seat of our federal government, the NCR presents a highly symbolic target with critical infrastructure that affects all Americans. These unique characteristics make the NCR more at risk to terrorist attack that almost any other area of the country. Unfortunately, many of the families affected on 9/11 know this to be true.

So does the Department of Homeland Security. Its internal assessment indicates that the NCR is a top target for future acts of terrorism. The Department identified 3,417 regional assets spread across eleven critical infrastructure sectors, placing the NCR in the top 25 percent of all eligible regions at-risk. By several unofficial accounts, the NCR is in the top 5 of all at-risk regions across the country, falling only behind New York City and other urban areas far more populated.

Whatever risk criteria are used, it is clear that the NCR confronts enormous challenges to protect its citizens, communities and the millions that visit our nation's capital.

The enormity of these challenges means that the NCR bears a disproportionate burden of security demands and costs. Since 9/11, the region has undertaken unprecedented efforts to recover from and prepare against future attacks. Through Fiscal Year 2005, these efforts were assisted by $167 million in Urban Area Security Initiative (UASI) grants from the Department of Homeland Security. However, substantial unmet needs still exist. Recently, the NCR has documented an additional $190 million in necessary security enhancements to meet the national priorities outlined in the National Preparedness Goal, the federal standard required of at-risk urban areas.

Despite the overwhelming evidence of the serious and persistent threat to this region, the Fiscal Year 2006 Homeland Security Grant Program for the NCR was slashed by almost half, from $77 million in FY2005 to $46 million in FY2006 – reducing the NCR's share of overall UASI funding from 9.4 percent to 6.5 percent. This year's UASI allocation falls drastically short of what the region's emergency preparedness professionals believe to be necessary to enable the NCR to prepare for, respond to and recover from a terrorist attack. This substantial reduction is dangerous for both the security of the NCR and the security of the nation.

The final FY2006 UASI allocation is based on a new, highly complex formula that few outside of the Department of Homeland Security can comprehend. In its first year, this new formula appears to have a established a paradoxical and unwise result – one where the highest at-risk targets, the NCR and New York City, have significantly reduced funding. One would think that this inverse relationship of risk and funding was precisely what the new formula was designed to avoid.

The Department offers a two-pronged justification for this outcome. First, it argues that the NCR and New York City have already received large investments under past allocations and therefore need less resources now. This is hardly relevant, and is contrary to the specific recommendation made by the 9/11 Commission that DHS make purely risk-based funding allocations. Unmitigated risk, not prior assistance levels, is the litmus test for DHS assistance. Homeland security assistance is not a general revenue sharing program; It is to enable the country's most at-risk regions to build the infrastructure to eliminate or substantially reduce these risks.

The Department's second justification – the "effectiveness" of the NCR application – is a bit more complicated. I applaud the Department's effort to ensure that homeland security grants will be spent effectively. However, I have serious reservations that the NCR application warranted a 40 percent cut from last year's funding. Months of evaluation, planning and analysis were dedicated by NCR professionals to develop a series of proposals to meet still unmitigated risks this region faces, as well as the many priorities established by the National Preparedness Goal. Indeed, these professionals identified proposals totally over $190 million within their application. These projects were needed to further reduce risks the NCR continues to face. It is inconceivable that DHS saw fit to fund less than 25 percent of their demonstrated needs.

The outcome demonstrates that the DHS grant formula, at least as it relates to the NCR and New York City, must be retooled. These two regions, joined by the common devastation of the 9/11 attacks, represent the heart of our federal government and the center of the world's financial commerce. An attack on these regions is not merely an attack against significant physical assets – it's an attack that strikes at the soul of our nation, with repercussions far greater than just the damage inflicted to infrastructures. Since the new DHS formula obviously fails to include these very significant intangible assets of the NCR and New York City, I urge the Department to revise its formula so it fully accounts for the risks to these assets. One option, I note, would be to set aside a pool of funding for these regions before the determining the allocations to other areas; another would be to limit the number of areas eligible for UASI grants.

Finally, I am concerned that the Department was unable to approach many of the most affected Members of Congress before it released its allocations publicly. The NCR has always been one of the few regions that routinely hold meetings to address security concerns and it actively seeks coordination with DHS. These cooperative efforts have helped overcome many of the daunting challenges we face. Many of us were taken aback when the Department announced the allocation without providing advanced notice for the more adversely impacted regions. If changes to federal statutes or appropriations report language are needed to avoid this predicament from happening again, I urge the Department to pursue them.

I look forward to working further with the Department on these concerns, as well as with the regional officials that have spent months deliberating over how best to secure the NCR. The lessons we've learned from this round of security grant funding can only help us all improve upon the system so that our region is ready in advance of any future terrorist attack.

Mr. Chairman, thank you for this opportunity.

Chairman TOM DAVIS. Thank you very much.

Mr. Van Hollen.

Mr. VAN HOLLEN. Thank you, Mr. Chairman.

Thank you for holding the hearing today on a very important issue. As some of my colleagues said, this does raise the overall question of our national priorities. We hear a lot of rhetoric from the White House and the administration about Homeland Security. It is a top priority. Yet when you look at the resources devoted to it, you find out that, hey, look what happened here?

One of the reasons we have seen cuts in Washington, DC, and New York is because we don't have the resources this year that we had last year. If it is the administration's position now that the threat to the homeland has reduced that much, it would be interesting to hear from them. Otherwise, it would be nice to have the resources match the rhetoric.

We are often told that we shouldn't prepare for the last war, and that is absolutely true. You got to always be thinking about new possible threats. On the other hand, I don't think there is anybody with a shred of common sense that can tell you that, going forward, the Nation's Capital and New York City don't remain top targets of al Qaeda and any other group, terrorist group, seeking to do our Nation harm. Clearly, they remain at the top of that list.

Now, many of us have been concerned for some time about the degree of attention the Department of Homeland Security has been paying to the Nation's Capital. We had a back and forth a little while ago about the downgrading of the position of the Office of the National Capital Region Coordinator within the Department of Homeland Security who used to report directly to the Secretary. The Secretary has now apparently determined that he will be buried within a couple of layers of bureaucracy. We haven't gotten any response to the complaints many of my colleagues and I wrote to the Secretary with respect to that downgrading. Now it appears that downgrading is not just with respect to the regional coordinating position, but with respect to the resources that are devoted to the National Capital area.

Now, this is an important hearing because we are going to be able to find out a little bit more why these decisions were made, because there are conflicting comments in the media. Some media reports say that the Homeland Security evaluation that New York City and the Washington, DC, area were given lower amounts because they were inadequate, that they were not well thought out and substantial. Yet there are other comments from representatives, including some on the panel from the Department of Homeland Security, that said that is not the reason they were downgraded. So I think it is a mystery to many people why these areas received less money.

I would just close by pointing out that the Washington, DC, area, as we all know, is a multi-jurisdictional area. You have two States, Virginia and Maryland. You have the Nation's Capital. Within the two States, you have lots of jurisdictions. Coordinating that area is a major challenge. To have the one source of money that really helps pull all those different jurisdictions together and working as one and pulling them in the same direction, to have that source of money cut sends I believe absolutely the wrong signal at the wrong time. There are lots of good ongoing efforts in the regions to try to bring everybody together. I think they can do even better, but they can't do even better with a lot less than they are getting now.

So this is a great mystery to many of us, how this decision was reached. Thank you, Mr. Chairman, for holding this hearing. Hopefully, we will get some answers.

Chairman TOM DAVIS. Thank you very much.

Mr. Ruppersberger.

Mr. RUPPERSBERGER. Thank you, Mr. Chairman, for having this.

Mr. Moran, I am glad you are here to make an argument.

First, I represent Maryland's Second Congressional District, includes the Baltimore region, part of the National Capital area. Also in my district is the port of Baltimore, NSA, BWI Airport, Fort Meade, Aberdeen Proving Ground and the sector of Baltimore which is a Coast Guard shipyard.

In 2006, the Department, as we know, that is why we are having this hearing today, the Department of Homeland Security cut National Capital Regional funding 40 percent from the previous year. They also cut $18 million in Homeland Security grants from the State of Maryland from the previous year.

Now, it is of great concern to me that DHS would cut counterterrorism money for New York City, the Washington region, which together have been 100 percent of al Qaeda's terrorist attacks on American soil. Each jurisdiction has been cut, based on my investigation, by, again, at least 40 percent.

Now, it is also of great concern to me that DHS has shown by their actions that the National Capital Region is a low-risk city. That is preposterous. When you compare the State of Maryland, which I represent, with nearly 5.5 million people, the per capita spending for DHS dollars is $4.5 per person. That means each person in Maryland gets $4.5 that could be allocated.

You look at Wyoming. Wyoming received nearly $16 per person in DHS funding per capita. Wyoming gets more than 15 times as many DHS dollars than they have people. $7.6 million and only 493,000 people who live in Wyoming.

I like Wyoming. The Grand Tetons is one of our most magnificent rural areas in the country. I have been there. I observed grizzly bears and other wild animals. Based on this new funding system, it seems that DHS is protecting the grizzly bears in Wyoming over the people who live and work around Fort Meade and NSA and other parts of this country. I choose people over bears.

Now, I am willing to listen to DHS's argument, and they are on the hot seat, as they should be. But I feel very strongly that we need to focus on our high-priority targets. I am a Member of the House Select Intelligence Committee. We focus on terrorism and al Qaeda all the time. I am just very, very concerned of DHS's priorities as it exists here today.

If DHS has a better understanding of where the risk is, why was the funding for such densely populated urban areas like Washington, DC, New York and Baltimore cut so much? The new DHS system which supposedly balances risk and effectiveness places more emphasis—and this is what really concerns me—places more emphasis on the quality of the writing of the grant proposal than the actual risk a community faces. I want to repeat that because that seems to be the argument that certain jurisdictions might not have written the grant proposal the way DHS wanted it.

I agree that giving money to States without having a plan is putting the cart before the horse, and DHS should have been criticized for the amount of waste, of money that went to different jurisdictions, pork, for dog vests and everything else. That has to stop. I am sure that is why this system is coming here now. But giving Homeland Security dollars on the basis of the quality of how a proposal is written is trying to drive the cart without a horse.

If the State plans are being analyzed, where is the congressional oversight? I question whether a group of peer review experts should be telling first responders what projects have value and which don't and determining how much money they should receive. Have these peer review experts ever managed a Federal grant or Federal emergency? I don't know, because I don't know who they are. They haven't been before the committee. There is an argument about being classified. I am on the committee that is classified. So there needs to be oversight with respect to that.

This process is not acceptable, and I believe the new DHS system puts our most vulnerable cities at risk. I think the DHS has put the needs of a few over the needs of the most vulnerable. I will work with my colleagues to fix the DHS grant formula so it accurately protects America's most vulnerable assets, our families, our community and our Nation.

Thank you, Mr. Chairman.

Chairman TOM DAVIS. Thank you.

Mr. Mica.

Mr. MICA. Thank you, Mr. Chairman. I really don't have a lengthy opening statement, which I am sure many will appreciate. But I am actually looking forward to hearing from some of these folks. I have never seen a goofier list of priorities for receiving Department of Homeland Security grants than the one most recently provided us.

I just want to figure out how they figured out how to mess things up so badly, particularly addressing the needs of some of our most critical cities and regions, Washington and New York and others. I want to hear the explanations to the goofy decisions.

Thank you. I yield back.

Chairman TOM DAVIS. Thank you.

Ms. Watson.

Ms. WATSON. Thank you so much, Mr. Chairman. This hearing is very important to our oversight responsibilities, and I would like to thank all the witnesses that are here today for coming.

I believe homeland security manifested as guns, barriers and blind response protocol is a false sense of security. Homeland security should be about the health of the American people. It really is not about the land; it is about the people; the people's education and the proper expertise that can be brought to strategic management and preparedness.

You see, these alerts go high every day. What do you do? The people have no clue. The Department of Homeland Security was born in America's moment of crisis. Unfortunately, as a result, it seems the Department's management itself is in a perpetual moment of crisis.

We saw the most devastating results of this mismanagement during Hurricane Katrina last autumn. But the hidden disaster is the way that our local government and our first responders are blown about by the capricious winds of DHS's policy shifts. Now DHS has again changed the way it allocates grant money to the local governments that are expected to respond in a case of disaster.

I am still concerned that we have not fixed the problems that have plagued the way we allocate these funds since the beginning. It seems to me that the endless debate over allocation formula is a product of the President's failure to provide the resources necessary to secure the homeland, the people.

Rather than debate who is going to get the biggest piece of a shrinking pie, we need to base not just our formula but also our overall spending levels on the true threats and risks to America.

So I want to thank you, Mr. Chairman. I look forward to hearing from the witnesses about how they believe we can improve DHS's grantmaking policies, specifically in the National Capital area.

I yield back.

Chairman TOM DAVIS. Thank you. I think that will conclude opening statements and we can get to our panel. Members will have 7 days to submit opening statements for the record.

I recognize our very distinguished panel of witnesses.

We have the Honorable George Foresman, the Undersecretary for Preparedness, Department of Homeland Security; there is Mr. Edward Reiskin, Deputy Mayor for Public Safety and Justice; Mr. Robert Crouch, the Assistant to the Governor of the Commonwealth of Virginia for Preparedness; Mr. Dennis Schrader, the Director of the Governor's Office of Homeland Security of the State of Maryland; and Mr. David Robertson, the executive director of the Metropolitan Washington Council of Governments.

It is our policy to swear all witnesses in before they testify, if you would rise and raise your right hands.

[Witnesses sworn.]

Chairman TOM DAVIS. Mr. Foresman, we will start with you and move on down the line. Thank you.

STATEMENTS OF GEORGE W. FORESMAN, UNDERSECRETARY FOR PREPAREDNESS, DEPARTMENT OF HOMELAND SECURITY; EDWARD D. REISKIN, DEPUTY MAYOR FOR PUBLIC SAFETY AND JUSTICE, GOVERNMENT OF THE DISTRICT OF COLUMBIA; ROBERT P. CROUCH, JR., ASSISTANT TO THE GOVERNOR FOR THE COMMONWEALTH PREPAREDNESS, COMMONWEALTH OF VIRGINIA; DENNIS R. SCHRADER, DIRECTOR, GOVERNOR'S OFFICE OF HOMELAND SECURITY, STATE OF MARYLAND; AND DAVID J. ROBERTSON, EXECUTIVE DIRECTOR, METROPOLITAN WASHINGTON COUNCIL OF GOVERNMENTS

STATEMENT OF GEORGE W. FORESMAN

Mr. FORESMAN. Mr. Chairman, thank you.

Ranking Member Waxman and members of the committee, thanks to each of you for the opportunity to appear today along with colleagues from the National Capital Region to discuss the

2006 Homeland Security Grant Program and specifically the Urban Area Security Initiative.

I would like to highlight several key issues for the committee today in advance of the very important discussions about reducing America's overall risk from terrorism. The discussion on funding should not be an issue of placing the safety and security of any one person, community or State in America ahead of another. Managing risk is a national responsibility. The process used by the Department this year reflects the desires of the administration, Congress and the American people to make our entire Nation safer and more secure. It is in that vein that I feel compelled to present to the committee some clear facts.

First, the amount of funding available for the Urban Area Security Initiative program has been reduced by $125 million compared to 2005. This does in fact represent a reduction of 14 percent. The State Homeland Security Grant Program has been reduced by $550 million compared to 2005, representing a reduction of approximately 50 percent.

Second, the Department, at congressional direction, implemented an approach this year that includes both risk analysis and effectiveness scoring. It is an approach that is supported by the administration and reflects the evolution of our post-September 11th approach to managing the risk of terrorism.

I would like to draw two excerpts from the House Homeland Security Appropriations Committee report for 2006: We are at a turning point in the methodology for administering the First Responder Grant Program. Historically, funds have been distributed based on minimum percentages and population.

A little later, the report language reads: ODP, now known as the Office of Grants and Training in the Department, will begin a new methodology for administering the First Responder Grant Program. Funding will be targeted based on threat and risk while targeting gaps in preparedness.

This guidance from the Congress has informed the development of the process that we used this year.

Third, the risk to our Nation is better understood than even 1 year ago. National risk is an umbrella that encompasses community and State risks together. No single community or State constitutes the sum of America's risk. Our adversaries have publicly stated their desire to kill and injure our citizens, wreck our economy and destabilize public confidence. We know that there are key facilities across America that, if attacked, could cause grave harm to the people inside and in the immediate vicinity and cause potentially devastating ripple affects across the entire Nation, chemical and nuclear power plants, key transportation, telecommunications and energy hubs, as well as financial centers, just to name a few.

In 2004, at the earliest stages of our national efforts, we had documented approximately 200 facilities nationwide. Last year, we had documented about 11,300. Today we have documented more than 260,000 nationwide. This documentation process is critical to understanding both the scope and diffuse nature of our national threat.

In addition, the post-September 11th efforts to fuse the work of the intelligence and law enforcement communities at all levels of government is producing a more accurate picture of suspicious activity. This is activity that may point to an impending plot against Americans, our economy and our way of life anywhere in our country. This information also strengthens our analysis and underscores mode of attack of the past may not be the mode of attack of the future.

The Department has adopted the lessons from the tragedy of September 11th. It is important to connect the dots, to understand our vulnerabilities, their consequences and the realities of potential threats all across the country. Together, these inform our risk analysis.

The Department of Homeland Security's job is to protect an entire Nation. We are using all available information to guide us in doing just that. Just as we know that putting all police resources in a single neighborhood after a robbery or murder will not reduce the risk of crime in an entire community, we also know that reducing an entire Nation's risk cannot be accomplished by focusing resources in any one area alone.

Fourth, and with what I just mentioned in mind, analyzing risk is neither absolute nor is it static. We have a much more accurate understanding of our entire Nation's risk. New York and the National Capital Region do not suddenly have less risk. Because of our better analysis, we know that the risk measures in some other urban areas are actually higher than previously assessed in relation to New York and the Washington region.

In some cases, the share of national risk for these other communities actually doubled or tripled as a result of our incredible analysis this year. This means that additional resources had to be applied to these areas to help address the better understood risk as part of a truly national approach. Two-thirds of funding to the 46 urban areas was based on these risk measures.

Let me be very clear: New York City and the Washington region continue to be at the top of our risk consideration. You will see a chart on the screen that shows the risk curve in the upper righthand corner and the funding curve in the lower lefthand corner. You can see that funding corresponds to risk. In fact, more than 45 percent of all funds go to just 5 of the 46 urban areas, New York City, the National Capital Region, Los Angeles, Long Beach, Chicago and the San Francisco Bay area.

It is also important to note that these 46 urban areas constitute approximate 85 percent of our national urban area risk. I should underscore that as our knowledge and information continue to evolve, relative risk rankings could change again in the future.

Fifth, we assessed the effectiveness of urban area applications this year relative to each other in support of meeting national priorities. These assessments also included review of how investments matched their respective State strategies and potential for continuing beyond the flow of Federal dollars.

The effectiveness score is not in any way a measure of how well urban areas are implementing programs, managing their resources, succeeding in keeping their citizens safe or how well they write their grant applications. It is simply a mechanism for promoting a national unified approach to the national threat of terrorism by encouraging efforts that meet program directions set forth by Congress and the administration.

Effectiveness accounted for one-third of the allocation decisions, and it is important to note that the maximum increase or reduction of funds based on these measures would have been at most 7 percent for any of the urban areas. The vast majority fell into the median, meaning there was very little change.

Finally, we continue to work on improving communication. The Department is responsible for addressing a national threat, and that means by working closely with many stakeholders, including our partners at the State and local level. Before this position, I spent nearly a quarter of a century in local and State public safety activities, and I clearly understand that funding decisions can send unintended messages.

The message here should be very clear: Members of the committee, managing America's risk requires a national approach that applies Federal resources wisely to supplement the work of State and local governments. The Secretary and I continue to balance the need for maximum transparency in the funding processes with the need to avoid publicly giving our enemies a roadmap to our national vulnerabilities.

We will continue to work closely with our partners at the State and local level here in the National Capital Region and with the Congress to ensure that we protect the entire Nation and that we provide clearer understanding of the progress that we are making on reducing America's risk from terrorism on numerous fronts.

Mr. Chairman, members of the committee, thank you for your attention and I look forward to your questions.

[The prepared statement of Mr. Forseman follows:]

Statement for the Record

George W. Foresman
Under Secretary for Preparedness
Department of Homeland Security

Before the House Government Reform Committee

June 15, 2006

Introduction

Chairman Davis, Ranking Member Waxman and Members of the Committee, thank you for the opportunity to appear before you today to discuss the Homeland Security Grant Program and specifically, concerns raised about the allocation process for the Urban Areas Security Initiative (UASI) funds.

There has been much debate and discussion during the past two weeks. Some of the information presented in public has been accurate and some has not. The debate itself is positive – it is welcome and necessary for the public to be engaged in discussion over homeland security priorities and funding.

One thing however is very clear: the discussion on funding should not be an issue of placing the safety and security of any one person, community or State in America ahead of another. This is very much about making our entire nation safer and more secure by managing risk in a way that lessens the vulnerability of the entire country.

The safety and security of each and every American lies at the core of the mission of the Department of Homeland Security (DHS), and it is a mission that the men and women of the Department take seriously.

However, a safer and more secure America is not an exclusive mission of the Department of Homeland Security. America's safety and security is a shared national responsibility. It is a mission that is shared among local, State and Federal agencies, the public and private sectors and the American people. In the context of terrorism, it requires an unprecedented mix of efforts – border and immigration controls, security in our ports, and airports and protection of critical assets and infrastructure, including transportation, communication, financial and energy. Homeland security is about managing risk for the entire nation based on a comprehensive national approach; it is about applying limited resources most effectively based on our understanding of America's overall risk.

Let me be very clear, there is a critical distinction to be made: Threat is not synonymous with risk, nor is risk analysis synonymous with risk management, as I will discuss later.

There are many tools employed every day and in every way to keep our nation safer and more secure from the threat of terrorism and a host of other hazards and threats that comprise our national risk continuum. Today, I would like to focus on the Homeland Security Grant Program (HSGP).

The HSGP is the Department's primary means of homeland security assistance to the states and local communities, and it includes the State Homeland Security Program (SHSP), the Law Enforcement Terrorism Prevention Program (LETPP), and the Urban Areas Security Initiative (UASI), along with the Metropolitan Medical Response System and Citizen Corps Programs. As such, HSGP is one of the Department's most important and visible mechanisms to manage national strategic risk.

Today's testimony will focus on the method DHS utilized to evaluate the risk of terrorism to States, territories, and Urban Areas; the peer review process we employed to determine the expected effectiveness of proposed solutions, and ultimately, the risk management techniques we used to determine allocations for Fiscal Year (FY) 2006. I will go into great detail regarding how the Department strived to employ an objective, comprehensive, and fair process for allocating FY2006 HSGP grants to improve nationwide terrorism preparedness.

The debate about "who got how much" has overshadowed the more important discussion about the best way to use limited financial resources to increase America's security. We used an approach this year that expands our understanding of what constitutes risk while taking into account Congressional guidance encouraging our nation to move away from "reaction" to "strategic preparation."

As Secretary Chertoff said in recent remarks pertaining to this program,

"We cannot protect every single person at every moment in every place against every threat. What we have to do is manage the risk, and that means we have to evaluate consequence, vulnerability, and threat in order to determine what is the most cost-effective way of maximizing security."

The Department's grants programs have traditionally provided financial assistance to all 50 States, the District of Columbia, the Commonwealth of Puerto Rico, and the U.S. Territories. By the end of Fiscal Year (FY) 2006, states and localities will have received from DHS over $18 billion in assistance and direct support from the Department of Homeland Security since September 11, 2001. This does not account for the additional billions made available from the Departments of Health and Human Services and Justice.

The Department is making significant, important, and vital changes to HSGP, both with the analytic capabilities that support the program and the management techniques we use to determine allocations. And, as we have all seen from the reaction to our FY 2006 allocations, implementation of risk management will not necessarily be an easy or a popular shift. However, it is an important shift and one that we take seriously. We have and will continue to solicit feedback on our processes and are willing to listen to criticism and suggestions for improving our processes. With billions of dollars being allocated each year, this is a serious business - and we believe that healthy debate about risk management principles will only make these processes better and more transparent. Despite recent successes globally in the war on terror, America's security will be a marathon and not a sprint. We need an objective funding process that will sustain improvements for the long- term.

Today, I hope to articulate the following policy considerations:

1) The objectives of the Homeland Security Grants are to enhance capabilities to prevent, deter, respond to, and recover from acts of terrorism, to be allocated based on risks, threats, vulnerabilities, and unmet target capabilities. It is long-standing Administration

policy that the limited pool of Federal grant resources should be used to improve long-term capabilities that provide a maximum return on investment, instead of to finance day-to-day reoccurring local [first responder] operational costs.

2) The new DHS risk assessment process incorporates the tremendous increase in risk information, which has improved our understanding of both specific risks, relative risk and the distribution of risk across our entire nation.

3) In applying risk assessments to the grant process, the Preparedness Directorate has emphasized the principle of risk reduction, including the peer-review scoring, which assesses the likelihood that Federal resources will reduce long-term risk and address short falls in capability. The new allocation formula, based on risk and effectiveness, strives to provide an objective process that is flexible to account for improved informational awareness and future evolutions in threat, consequence, and vulnerability.

FY 2006 – A Transition Year

FY 2006 marks the first HSGP grant cycle in which the Interim National Preparedness Goal is in place to identify National Priorities and help focus local and state expenditures. This common planning framework, and the tools that support it, allows us individually as communities and states and collectively as a Nation to better understand how prepared we are, how prepared we need to be, and how we prioritize efforts to close the gap. The absence of this type of consistent preparedness target is at the forefront of many of our national shortcomings over the past 25 years. The Interim National Preparedness Goal demands that we focus attention on "raising the bar" of preparedness across the country to establish minimum capabilities and be prepared for the risks we face. This, along with measurement of risk, gives us an important management consideration for our grant programs.

Accordingly, the Department of Homeland Security has been aggressive in:

1) improving the risk analysis tools used to determine a National risk profile, so that we can target funding at higher risk locations, and

2) clarifying the risk management objectives for the HSGP, within the context of the National Preparedness Goal

This year we have also implemented another significant change in how funds under the HSGP are allocated. In previous years, States and Urban Areas knew their funding allocations prior to submitting grant applications. Based on substantial input from the national preparedness community Congress, and our focus on risk management, Department has moved towards a risk-based approach that incorporates a competitive element, to allocating funds for HSGP. This is a critical step in achieving a Homeland

Security Grants Program that emphasizes risk-informed grant making, increased accountability and is focused on maximizing the return on investment of grant funds.

Applying Risk

Key to this year's process is a much better understanding of our national risk. New York City and the National Capital Region (NCR) clearly remain at the top of the list of American urban areas with a risk from terrorism. We know this for several reasons, including the facts that they were the sites of the horrific attacks on September 11, 2001 and that there has been much open source reporting about them being targeted since that time. A key criticism of our national approach to addressing the terrorist threat pre September 11th is the fact that we as a nation were not adequately preparing for multi-faceted threats from our adversaries.

Our enemies still wish to inflict both physical and economic harm on the United States. The understanding of their intentions is informed by the work of our national intelligence and law enforcement apparatus, public statements of their leaders as well as materials acquired from a variety of open sources and reported extensively by the media. This combination has informed our national risk management strategies that include a better understanding of how attacks targeting specific sites can not only kill or injure people nearby, but can also cause significant disruption of our national economy and potentially undermine societal confidence.

This understanding is widely shared and underscores both the Administration's and Congress's desire to assess and categorize our national assets – things such as key transportation hubs, financial processing sites, nuclear power and chemical plants, priority communication and energy systems. These are sites that, if attacked, would have an extraordinary impact not only on the surrounding population and community, but in some cases, the nation as a whole. In the first year of this grant program we had categorized approximately 200 sites, in 2004 some 1700, in 2005 approximately 11,300. This year, we further expanded the number of sites to include many considered to be 'high risk' by the surrounding state and local jurisdiction, which brought the total number of sites in the analysis to over 260,000 sites.

With improved assessments, we are better prepared to protect critical infrastructure across the nation. We are better able to work with our colleagues in the intelligence and law enforcement communities. A major effort since 9-11 has been geared toward the fusion of intelligence and law enforcement reporting the so called "connecting of the dots". Through our partnerships with state and local authorities, we have a better understanding of the reporting of suspicious monitoring activities by local and state authorities that may constitute surveillance, as well as a more organized approach to including information from on-going terrorism related investigations. These together help inform our risk analysis.

Relative Risk

With the enhanced methodology and broader set of data inputs, we were able to capture a truer estimation of relative risk for all urban areas. The footprint within which we analyzed the risk to both assets as well as geographic areas and populations was broadened this year to more accurately reflect the regional context in which these jurisdictions operate and the critical infrastructure for which, in coordination with the private sector, public safety officials in these areas are responsible. We had better data about not just New York City and NCR, but about the entire country and across a broader range of sectors. As a result of these improvements, many areas' risk scores changed significantly, a reflection of a more sophisticated analytical approach to gauging the risk urban areas face relative to one another.

It is important to understand the downstream impact of these changes in relative risk. New York City and NCR suddenly have less risk in an absolute sense. New York City and NCR continue to be among the highest risk Urban Areas. However, the relative values for virtually all other candidates increased this year due to our better understanding of risk and how to analyze it. The relative differences among the higher risk candidates is what changed from last year to this year. Indeed, Urban Areas such as Los Angeles, Chicago, and Houston saw their share of national risk relative to New York City and NCR increase considerably, in some cases doubling or tripling compared to previous analysis. These changes in relative risk were key drivers in the changes in funding allocations.

Assessing Risk

In past years, DHS' risk analysis was largely driven by population size and density. But over time we have been able to develop enhanced techniques to analyze risk. In FY 2006, the risk analysis considered three primary components: Threat, Vulnerability, and Consequence. The Threat component represents an adversary's intent to attack a specific target and its potential capability to execute the attack; the Vulnerability component embodies the susceptibility to an adversary's attack and the likelihood that it will achieve an impact; and the Consequence component measures the possible impact from such an attack.

Specifically, in the UASI program, which is part of the HSGP, the bulk of these funds have gone to those urban areas that are generally recognized to be the highest risk in the Nation. In fact 45 percent of the funding has gone to just five of the urban areas— New York City, the National Capital Region, Los Angeles/Long Beach, Chicago, and the San Francisco Bay Area.

HSGP and Risk Management

I would like to explain how we analyzed risk for determining the 2006 grant funding.

The Department of Homeland Security has many risk management resources at its disposal—people, technology, and funding are just a few. The HSGP is among the most valuable of these tools because it allows us to partner with our States, Territories and Urban Areas and First Responder communities, and support national preparedness goals.

The Administration, Congress, State and Local stakeholders, first responder organizations, and industry groups have called for more risk management approaches to inform homeland security grant allocations. There has been a clear recognition that our national approach requires that we apply federal funding resources in a way that maximizes resources to benefit all Americans.

In our effort to improve our methods for risk management of the terrorist threat we considered several key factors.

1. Ultimately, it is the States, Urban Areas and Territories that own the risk in their respective areas, and they must make investments locally that will build needed capabilities and address identified risk. DHS's risk management job is to provide them guidance, and within available resources, financial assistance to make these investments. In this program, we have been directed to invest in initiatives that promote unity of effort at the community, regional, state, and national levels. They must continue to provide tangible benefits beyond the flow of Federal dollars.

2. When managing risks, we must rely on analysis of risk to inform our management process, but be cognizant of the inherent uncertainty of this analysis. Consider this definition of risk analysis from the Society for Risk Analysis:

   "Risk analysis uses observations about what we know, to make predictions about what we don't know."

   I think this sums up risk analysis in the context of homeland security quite nicely. We have carefully considered the factors that experts believe lead to risk, and we have confidence in our approach. But we are realists and we understand that risk in the terrorist context is new, constantly changing, and lacks the measuring history of data flow found in other hazards.

   Terrorist threat cannot be predicted with the reliability of hurricanes or floods, or mechanical failures. No matter how much we invest in scientists and algorithms, we cannot measure terrorism risk in an absolute sense. Therefore, we emphasize building capabilities to manage risk nationwide based on the best estimations possible. Our profile is built on an analysis of relative risk—based on what is known.

3. Risk Analysis DOES NOT EQUAL Risk Management. In fact, the Society for Risk Analysis definition makes this point better than I can:

Risk analysis seeks to inform, not to dictate, the complex and difficult choices among possible measures to mitigate risks.

As this indicates, the risk analysis is only one input to the risk management process that should be considered for Homeland Security. In any risk context, risk management typically involves considerations beyond the quantifiable analysis. Risk management includes many other considerations such as management objectives, fiscal constraints, one's ability to actually impact the risks one faces, and the strategy that best serves our overall national interests. The primary risk management objective of the HSGP is to: raise the bar of preparedness across the at-risk states, territories and Urban Areas as part of an interdependent national effort by directing funds to areas of greatest risk and need.

These two objectives announced by Congress require the Department to balance the desire to focus resources on areas at relatively greater risk, with the desire to promote use of federal resources for strong solutions that "raise the bar" of national preparedness and address national risk.

Thus, common sense dictates that managing risk through the HSGP program involves much more than just distributing dollars in proportion to the relative risk data that we generate each year. Rather, it is viewed as a means for reducing risk and promoting national objectives.

Promoting Effectiveness

For FY 2006, States and Urban Areas submitted grant applications, called Investment Justifications, to formally request FY 2006 HSGP funding in support of their strategies and related program planning documents. These applications were reviewed through an intensive peer review process. The FY 2006, competitive grant process to allocate funds to States and Urban Areas was based on two factors:

1) The relative risk to assets and populations within the eligible applicant's geographic area, and

2) The anticipated effectiveness of the individual investments comprising the Investment Justification, in aligning to the Interim National Preparedness Goal and addressing the identified homeland security needs of each applicant.

Finding the right balance between these two factors is the central risk management challenge. It requires us to conduct extensive analysis of relative need and risk, thoroughly review applications, and rigorously analyze the potential effectiveness of the grant funds. The Department of Homeland Security conducted an unprecedented amount of analysis to arrive at decisions about grants funding. We took into consideration alignment with other national policy initiatives and statute objectives, as well as ensuring consistency of approach both over time and between the HSGP programs.

The major considerations of project requests were the following:

- Relevance – Connection to the National Priorities, Target Capabilities List, State/Urban Area Homeland Security Strategy goals and objectives, and the Enhancement Plan.
- Regionalization – Coordination of preparedness activities across jurisdictional boundaries by spreading costs, pooling resources, sharing risk, and increasing the value of their preparedness investments.
- Impact - The effect that the investment will have on addressing threats, vulnerabilities, and/or consequences of catastrophic events.
- Sustainability – The ability to sustain a target capability once the benefits of an investment are achieved through identification of funding sources that can be used beyond the current grant period.
- Implementation Approach – The appropriate resources and tools are (or will be) in place to manage the Investment, address priorities, and deliver results.

The peer review process provides a significant incentive for States and Urban Areas to spend the limited pool of Federal resources on projects that will provide a meaningful return on investment and a lasting impact on reducing the risks of terrorism.

FY06 HSGP Allocation Formulations

Given the considerations I have just outlined, our task in transforming the HSGP this year was to build an allocation process that balanced our existing analysis of relative risk with our belief in the anticipated effectiveness of the Investment Justifications provided by the States and Urban Areas.

Risk analysis

As noted, DHS defines risk by three principal variables: Threat, or the likelihood of a type of attack that might be attempted, vulnerability, or the likelihood that an attacker would succeed with a particular attack type, and consequence, or the potential impact of a particular attack. The risk model used as input to the HSGP process includes both asset-based and geographically-based terrorist risk calculations. DHS combines these complementary risk calculations to produce an estimate of the relative risk of terrorism faced by a given area.

The asset-based approach uses strategic threat estimates from the Intelligence Community of an adversary's intent and capability to attack different types of assets (such as chemical plants, stadiums, and commercial airports) using different attack methods. DHS analyzes the vulnerability of each asset type relative to each attack method to determine the forms of attack most likely to be successful. Additionally, DHS estimates the consequences that a successful attack would have on each asset type, including human health, economic, strategic mission, and psychological impacts. This analysis yields a relative risk estimate for each asset type, which DHS applies to a given demographic area, based on the number of each asset type present within that area.

The geographic-based approach allows DHS to consider general characteristics of a geographic area mostly independent of the assets that exist within that area. First, DHS evaluates reported threats, law enforcement activity, and suspicious incidents reported during the evaluation period. Next, DHS considers vulnerability factors for each geographic area, such as the area's proximity to international border. Lastly, DHS estimates the potential consequences of an attack on that area, including human health, economy, strategic mission, and psychological impacts.

DHS's ability to analyze risks to the Nation is improving each year in both breadth and sophistication. Despite the known limitations of the Department's analysis, the results confirm two fairly intuitive points:

1) The majority of the risk is contained in a handful of locations throughout the country. This is the argument so strenuously made by that handful of localities. However,

2) There are risks to other urban areas that we have begun to assess more accurately. These areas have previously received relatively small amounts of grant funding. The HSGP risk analysis considered much more than the final number of cities that made the Urban Area list. Those that made the list did so because they had a level of risk. In this case, the urban areas under UASI contain 85% of our national urban area risk. Attachment A reflects both the funding and risk curve and you can see these correspond.

Given these two results, and drawing on intuition and common sense, it seems reasonable that while we must fortify higher-risk locations, we cannot ignore the risks in the other locations.

Anticipated Effectiveness

States and Urban areas each submitted up to 15 investments for consideration. These investments were submitted with an Investment Justification, which allowed them to describe specific funding and implementation approaches that would help achieve initiatives outlined in the Statewide Program and Capability Enhancement Plan. This plan developed in the Fall of 2005 establishes how Urban Areas and States will work to develop their individual capabilities as part of a broader national effort. The Investment Justification allowed the States and Urban Areas to request funding for allocation to their near-term priorities, consistent with the National Priorities articulated in the Interim National Preparedness Goal

The effectiveness review is a method to evaluate a state or Urban Area proposal in relation to others submitted and against the grant program criteria provided. It is not, I repeat it is not an evaluation of how well an initiative is or is not performing in a particular State or Urban Area. This element, added with Congressional direction and support, is designed to encourage uses of funds in accordance with pre-announced program guidelines and that will both enhance community, state and national preparedness beyond a grant period.

As we are not allocating funding to specific investments, our risk management objective was to determine the "anticipated effectiveness" of the investments contained in the Investment Justification. To do this, DHS convened a panel of a cross section of representatives from States, Territories, and Urban Areas, and from a variety of Homeland Security and Emergency Management disciplines.

States and Urban Areas sent high ranking officials to be reviewers; for example, three States sent their most senior Homeland Security Directors. From the Fire and Rescue community, an Assistant Deputy Fire Chief, Battalion Chief, Fire Operations Chief, and a Fire Emergency Management and Communications Chief participated, from Law Enforcement, an Assistant Chief of Police, Captain of a Sheriff's Department, Commander of a Special Response Team, and a Lieutenant from a Homeland Security and Tactical Operations. All used their knowledge and experience to evaluate the anticipated effectiveness of proposed solutions from their peers. These examples are only a subset of the vast experience of peer reviewers who participated in the HSGP process.

Peer review panels were made up of reviewers from varied backgrounds and experience - and to avoid potential conflicts of interest - diversity was emphasized. Each panel included a balance of representation from each region (Eastern, Central, and Western). The peer review panels reviewed and scored each individual Investment included in the Investment Justification as well as the Investment Justification submission in its entirety. The peer review panels also reviewed the Enhancement Plan to ensure alignment among Initiatives from the Enhancement Plan with proposed Investments.

As expected, the scores for the individual investments followed a distribution from very low to very high, with the majority of scores falling in the mid-range.

The anticipated effectiveness score for each State, Territory, or Urban Area was calculated by combining the weighted average of the individual Investment scores and the overall Investment Justification score. This combination gives a broad measure of both the overall vision of the State, Territory, or Urban Area homeland security program and the specific intent of the proposed investments.

Allocation Model

To support the management objectives of HSGP, we investigated several allocation techniques, and ultimately arrived at two management decisions. First, we gave particular attention to the analysis for New York City and the National Capital Region to ensure that the allocation process optimally accounts for their risk information and infrastructure assets. In addition, we selected a two-by-two matrix approach that allows us to evaluate Investment Justifications based on the Relative Risk to the Applicant vs. the anticipated Effectiveness of the Investment Justification submitted by that applicant.

This two-by-two matrix approach provided us with the following benefits:

- It allowed us to assemble a picture of the challenge recognizing that the two factors we value: Relative Risk and anticipated Effectiveness are distinct and not inherently correlated

- It gave us a relatively simple lens through which to view the decision space as policy makers, while still allowing a known model to drive final allocations.

To generate final HSGP allocations, we assembled two of these matrices: one for States and Territories subject to SHSP and LETPP dollars, and one for Urban Areas subject to UASI dollars. The matrices worked the same. Each applicant was plotted in the matrix by using their relative risk score and their Investment Justification Effectiveness rating.

Once plotted in the matrix, each applicant fell into one of four quadrants:

- Quadrant 1: higher relative risk/higher anticipated effectiveness

- Quadrant 2: higher relative risk/lower anticipated effectiveness

- Quadrant 3: lower relative risk/higher anticipated effectiveness

- Quadrant 4: lower relative risk/lower anticipated effectiveness

Once allocations were determined for each of the four quadrants, final dollar allocations were determined. For that, Relative Risk was weighted two-thirds and anticipated effectiveness was weighted one-third to emphasize the risk-based nature of the programs while recognizing strong program solutions. Using our analytic model, we generated the final allocation results you have seen, and which are illustrated by the chart below.

FY 2006 UASI Allocations

Urban Area	Allocation
AZ - Phoenix Area*	$3,920,000
CA - Anaheim/Santa Ana Area	$11,980,000
CA - Bay Area	$28,320,000
CA - Los Angeles/Long Beach Area	$80,610,000
CA - Sacramento Area*	$7,390,000
CA - San Diego Area*	$7,990,000
CO - Denver Area	$4,380,000
DC - National Capital Region	$46,470,000
FL - Ft. Lauderdale Area	$9,980,000
FL - Jacksonville Area	$9,270,000
FL - Miami Area	$15,980,000
FL - Orlando Area	$9,440,000
FL - Tampa Area*	$8,800,000
GA - Atlanta Area	$18,660,000
HI - Honolulu Area	$4,760,000
IL - Chicago Area	$52,260,000
IN - Indianapolis Area	$4,370,000
KY - Louisville Area*	$8,520,000
LA - Baton Rouge Area*	$3,740,000
LA - New Orleans Area	$4,690,000
MA - Boston Area	$18,210,000
MD - Baltimore	$9,670,000
MI - Detroit	$18,630,000
MN - Twin Cities Area	$4,310,000
MO - Kansas City Area	$9,240,000
MO - St. Louis Area	$9,200,000
NC - Charlotte Area	$8,970,000
NE - Omaha Area*	$8,330,000
NJ - Jersey City/Newark Area	$34,330,000
NV - Las Vegas Area*	$7,750,000
NY - Buffalo Area*	$3,710,000
NY - New York City	$124,450,000
OH - Cincinnati Area	$4,660,000
OH - Cleveland Area	$4,730,000
OH - Columbus Area	$4,320,000
OH - Toledo Area*	$3,850,000
OK - Oklahoma City Area*	$4,102,000
OR - Portland Area	$9,360,000
PA - Philadelphia Area	$19,520,000
PA - Pittsburgh Area	$4,870,000
TN - Memphis Area	$4,200,000
TX - Dallas/Fort Worth/Arlington Area	$13,830,000
TX - Houston Area	$16,670,000
TX - San Antonio Area	$4,460,000
WA - Seattle Area	$9,150,000
WI - Milwaukee Area	$8,570,000

*Sustainment Urban Area

The allocation process used this year to distribute the nearly $711 million in UASI funding, $125 million less than FY 2005 (overall HSGP funding was reduced $343 million below the President's request), to 46 metropolitan areas was structured to take into account both the risk and effectiveness of the proposed investments.

National Capital Region

The National Capital Region has received, on average, approximately 8 percent of all funding through the Urban Areas Security Initiative since the program's inception, and has received almost $214 million overall from the UASI program since 2003. Over that period, the NCR urban area has received third highest amount of grant funding from the UASI program, behind only New York City and Los Angeles/Long Beach Urban Areas.

The relatively high risk ranking played a major factor in the NCR receiving 7 percent of the total UASI funds available this year, nearly $46.5 million, and the allocation is clearly consistent with previous, annual allocation percentages.

As we look at investing Federal dollars, within the National Capital Region or elsewhere, we are seeking investments that promise to increase the overall capability of a region through funding such things as equipment and specialized training. Washington, DC, and its partners have worked hard in this area. However, we must also ensure that resources are also available to enable other at-risk communities to enhance their preparedness.

We must also consider the unique resources available to the National Capital Region through the permanent station of Federal operational resources that supplement what is being done by local and state officials. This includes air patrols, Federal law enforcement agents and other specialized federal response teams whose vigilance and capability may not be quickly available to other American urban areas. Together, these assets contribute to an integrated network that protects the National Capital Region.

Mr. Chairman it is essential to recognize that the grants process will only be successful if it is a nationwide effort to manage risk.

New York City

Let me turn briefly to the New York region. NYC remains the highest-ranked city for relative risk; of the more than 260,000 assets considered in the risk analysis process, nearly 7,000 came from New York City alone. However, due to the increase in information in our analysis and our better understanding of risk in regional areas, the "lead" that NYC had over other urban areas is smaller than it has been in past years. In simple practical terms, this means that there are very large UASI areas out there whose relative level of risk has "gotten closer" to that of NYC.

Since the creation of the Urban Area Security Initiative (UASI) program, New York City has received approximately 19% of the program's total grant funding, or more than twice the amount of the second largest recipient, even though the program now covers dozens of American cities. The next largest recipient, Los Angeles, has received approximately 8% of the funds awarded through the program.

HSGP Guidance

Prior to the release of the HSGP guidance, DHS provided extensive assistance to States and local governments in their development of updated Homeland Security Strategies and the Capability Enhancement Plans, which link investment planning to the National Priorities outlined in the Interim National Preparedness Goal. This guidance for the development of Enhancement Plans was a critical precursor to the development of successful Investment Justifications that meet the criteria assessed by the Peer Review Panel during the HSGP application process.

Between the time that the FY2006 Homeland Security Grant Program (HSGP) guidance was released on December 2, 2005, and the application due date of March 2, 2006, the DHS Grants and Training (G&T) Preparedness Officers for both the State of New York (NY) and the District of Columbia (DC) had frequent contact with NY and its Urban Areas, and DC and the National Capital Region (NCR) Urban Area. The officers were available to answer technical questions regarding the process. Due to the competitive nature of the application process, G&T staff members were not able to discuss or offer advice regarding specific program or budget proposals that may unfairly benefit one application over another.

G&T provided technical assistance to assist with the Program and Capability Review (PCR), which was the core planning process each State was required to conduct prior to submitting proposals. The PCR justified how any FY 2006 funds would be invested. Approximately 34 representatives from NY State, to include representatives from both the New York City and Buffalo Urban Areas, participated in the PCR technical assistance on November 30, 2005. Approximately 65 representatives from DC and the NCR Urban Area participated in their PCR technical assistance on January 5, 2006. Both of these sessions stressed the need to emphasize broad regionalization and include additional stakeholders, such as other local regions and the private sector, in the program planning process.

In addition to the formal PCR technical assistance deliveries, G&T Preparedness Officers had frequent, often daily, contact with the NY and NCR Urban Areas. As an example, the New York Preparedness Officer attended the NYC Urban Area Working Group meetings on a monthly basis, and a special meeting regarding the PCR process was held on November 28, 2005, for the NCR Urban Area Senior Policy Group. Representatives from DC and the NCR participated in the pilot development of the PCR technical assistance program on November 4, 2005, and served on the pilot working group to assist in shaping the PCR technical assistance offering. Feedback provided during the pilot was used to refine the design and materials prior to deployment to States and Territories across the Nation.

Conclusion

Mr. Chairman it is essential to recognize the distinction between risk and threat. Although threat is a large component of risk, *risk does not equal threat,* but considers it along with vulnerability and consequences. Likewise, *risk analysis informs, but does not equal risk management.* We now have a much better understanding of nationwide risk then we have in the past, along with the ability to evaluate risk mitigation strategies. As a result we now have a dynamic process for managing risk that reflects the Nation's priorities. We have come a long way in our understanding of risk and as we learn we will continue to improve this still evolving process.

Managing risk is a national responsibility. We would not be acting responsibly if we simply looked at each individual state or Urban Area as its own entity in making risk-based decisions. America's security requires a comprehensive approach and the federal government has an obligation to protect the entire nation. We must take steps necessary to ensure that all of our high risk areas increase their levels of capability. The grants allocation process is not about making Omaha, or Chicago, or Washington D.C. safe and secure—it's about making America safe and secure.

We must remember that providing grants to the states and Urban Areas is just one aspect of managing risk. Whether it's through border security, ensuring the security of nuclear plants, food storage facilities, financial centers across the country or cracking down on illegal immigrants, what we do in one area of the country will make a difference everywhere else.

We must work together to close gaps in America's security. We know that terrorists are working hard to exploit gaps in our efforts and the American people deserve no less than our very best effort to thwart those who would do us harm. I would like to thank the committee for its time today and I appreciate this opportunity to bring further transparency on this process.

Chairman TOM DAVIS. Thank you, Mr. Foresman.
Mr. Reiskin, thanks for being with us.

STATEMENT OF EDWARD D. REISKIN

Mr. REISKIN. Good morning, Mr. Chairman, Congresswoman Norton, members of the delegation and committee.

My name is Ed Reiskin, I am the Deputy Mayor for Public Safety and Justice for the District of Columbia, and I have the day-to-day responsibility for Homeland Security in the District. I am pleased to have this opportunity to be here today to talk about this very important issue for the District and the region.

Mr. Crouch from Virginia, Mr. Schrader from Maryland and I work together collectively and collaboratively on a day-to-day basis to improve the safety and security of the region, and that is a responsibility that we take very seriously, and it is in that spirit that we are jointly presenting our testimony today.

I will discuss our application, the risk and effectiveness assessment of it and the impacts to the District. My colleagues will discuss impacts to the region as well as impacts to pair to their respective States. If you are following in the written testimony, the State impacts are in the last pages of the testimony.

With regard to the application sample, some of the documents which are in front of me, significant effort by experienced practitioners from fire chiefs to health directors to communication experts, volunteer managers, hospital executives, many of whom I am pleased to see are in the room today, and many other stakeholders work toward the regional security of the homeland and worked together jointly to develop the fiscal year 2006 funding applications.

With our understanding of the fiscal climate and the Department's shift to a more risk-based approach, we were expecting an award somewhere in the area of $100 million to $120 million. To be conservative, we developed an application for $188 million, which was a prioritization of over $250 million of identified need that represented an amount we felt comfortable we would be able to execute. The breakdown of the areas that make up that $188 million are in your written testimony.

As many of you have mentioned, the Nation's funding for the Urban Area Security Initiative was reduced by 50 percent. We were reduced by 40 percent, yielding a $46.5 million award, which was, needless to say, considerably lower than we had expected.

With regard to risk and effectiveness, as many of you have said, we know we are a high-risk area. We understand through a press release that DHS found our region to be in the 97th percentile in terms of risk, and with risk as two-thirds of the equation for determining the allotments, it is hard for us to understand the disproportionate reduction that we seem to have received.

Absent any specific feedback from the Department, which we understand will be forthcoming, we can't speak to the effectiveness analysis of our proposal, although it was rated at or above average in each of their rating categories for each of the investment areas.

What we do know is that some of the most experienced responders and planners, many of which are in the room, people who did respond to September 11th, who have responded to anthrax, were the ones developing the content for this application. So we don't understand how peers from across the country could have somehow found this application to be lacking.

We have built significant capability in the region over the last 5 years that we are proud of and that should give comfort to those who live here, work here and visit here. That capability will not generally diminish as a result of this or any other award. But preparedness is a dynamic and complex process, and we have significant unmet need that remains, and we will be unable to meet all of it in this grant cycle. We can't speak to the levels of need or risk elsewhere in this country, but we do know that ours is high.

With respect to the District of Columbia, as many of you mentioned, we are the seat the Federal Government, home of the Capitol, the White House, the Supreme Court, the Department of Homeland Security and its national operations center, the FBI headquarters, the Washington Monument, 20 million visitors and countless other national icons and critical Federal functions. Yet the Department of Homeland Security determined that we face less risk than 75 percent of the Nation's States and territories.

The region was found to be in the top 25 percent of risk, but the city that serves at its core was in the bottom 25 percent. State Homeland Security grant funding was cut 50 percent nationally compared to last year. The District of Columbia's share was cut more than 53 percent. Only American Samoa, the Northern Marianas Islands, the Virgin Islands and Guam received less State Homeland Security grant funding than the District of Columbia.

Between State and law enforcement grant programs, from an identified need of over $37 million, we submitted an application for $21.5 million and were awarded $7.4 million. Our application included the following, which directly support national priority capabilities: $3.7 million for planning, training and exercising in areas such as national incident management system, hazardous materials response, continuity of operations planning; $1.6 million for enhanced preparedness, focusing especially on special needs populations, schools and businesses; funds for mass care, $1 million for critical infrastructure protection, there is no question we have a significant amount of critical infrastructure here; $3.4 million for intelligence fusion, following the guidelines of the Federal Government, both Homeland Security and justice; $2.4 million to continue our work in interoperable communications, including the dedicated wireless public safety network that is currently used to support events by both District and Federal agency users; $1.5 million for response to chemical, biological, radiological, nuclear weapons of mass destruction attacks that the District may face; $5 million for law enforcement response to raise our capabilities for our harbor patrol, our emergency response teams, our air support unit and others to Tier I as required by Homeland Security Presidential Directive 8.

There is obviously a lot more in the applications, and the bottom line is we won't be able to complete all of those things as we proposed.

I do want to close by saying we fully support a risk-based approach, and we support and acknowledge the Department of Homeland Security's efforts to make this process more objective and more transparent, which I do believe it has been. However, we have to question the outcomes, which don't seem to square with professional or common sense.

We are confident that the lessons learned from this year's process will ensure that next year's is better. With that, I will turn to my colleague from Virginia.

[The prepared statement of Mr. Reiskin follows:]

Governments of
the District of Columbia,
the Commonwealth of Virginia,
and the State of Maryland

National Capital Region's Homeland Security Senior Policy Group

Joint Testimony of
Edward D. Reiskin, Deputy Mayor for Public Safety and Justice for the District of Columbia
Robert P. Crouch, Jr., Assistant to the Governor of Virginia for Commonwealth Preparedness
Dennis R. Schrader, Director of the Governor's Office of Homeland Security for the State of Maryland

Regional Insecurity: DHS Grants to the National Capital Area

Committee on Government Reform
Representative Tom Davis, Chair
Representative Henry Waxman, Ranking Member

June 15, 2006

Room 2154
Rayburn House Office Building
Washington, DC 20510
10:00 A.M.

Testimony of the National Capital Region
Committee on Government Reform
Public Hearing on Homeland Security Grants to the National Capital Region
June 15, 2006

Mr. Chairman, Mr. Ranking Member and members of the Committee thank you for the opportunity to appear today to provide perspective regarding the homeland security funding decisions and the associated impacts in the National Capital Region (NCR)[1]. We have submitted our joint written testimony for the record. Consistent with the collaborative nature of the NCR, we opted for joint written testimony. We share goals, ideals, funding concerns and, most importantly, an intense commitment to the safety and security of the NCR that transcend the political boundaries within the NCR. Moreover, since preparedness requires coordination across the region, we plan, train, exercise, and work together on these issues and develop our proposals for funding to the Department of Homeland Security jointly. We are also joined today by Dave Robertson, the Director of the Metropolitan Washington Council of Governments (MWCOG).

You asked us to answer the following three questions:

- What are the impacts associated with the FY06 Urban Area Security Initiative (UASI) funding allocation?
- What are the potential increased security risks to the region?
- What is our assessment of DHS' risk and effectiveness-based process used for reviewing FY06 homeland security grant applications?

THE GREATEST IMPACT IS THAT THE AWARD AMOUNT WILL SLOW IMPROVEMENT AND INTEGRATION OF INTERJURISDICTIONAL CAPABILITIES

NCR Funding History

The National Capital Region is defined as an Urban Area by the Department of Homeland Security, making it eligible for grant funds under the Urban Area Security Initiative program. In addition, Virginia, Maryland, and the District receive DHS grant funds as states through programs such as the State Homeland Security Grant and Law Enforcement Terrorism Prevention programs.

The following table outlines the funding received or to be received by the NCR under the UASI program.

---

[1] Title 10, United States Code, Section 2674 (f)(2) provides the following definition:
The term "National Capital Region" means the geographic area located within the boundaries of (A) the District of Columbia, (B) Montgomery and Prince Georges Counties in the State of Maryland, (C) Arlington, Fairfax, Loudoun, and Prince William Counties and the City of Alexandria in the Commonwealth of Virginia, and (D) all cities and other units of government within the geographic areas of such District, Counties, and City.

Testimony of the National Capital Region
Committee on Government Reform
Public Hearing on Homeland Security Grants to the National Capital Region
June 15, 2006

Table 1[2]

Grant	Effective Award Date	Period of Performance	Grant Award	Percentage of total UASI program
03 Urban Area Security Initiative Part I	12/30/2003	6/1/03 – 11/30/05	$18,081,000	10.3%
03 Urban Area Security Initiative Part II	12/30/2003	6/1/03 – 11/30/05	$42,409,851	
04 Urban Area Security Initiative	3/29/2004	12/1/03 – 5/30/06	$31,921,361	4.7%
05 Urban Area Security Initiative	3/1/2005	10/1/04 – 3/31/07	$77,500,000	9.1%
06 Urban Area Security Initiative	Awaiting the award	To be determined	$46,470,000	6.3%
Total:			**$ 216,382, 212**	**7.6%**

As outlined in Table 1, the Department of Homeland Security granted the NCR approximately $170 million dollars through the UASI program from FY 2003 through FY 2005, equal to roughly 8% of the total UASI program.

Application Process

An understanding of the process used to develop the FY 06 UASI application will help explain what the impacts to the region will be. State, local, federal, and non-government stakeholders from across the region worked throughout January and February of this year to develop the application, based on the *FY 2006 Homeland Security Grant Program – Program Guidance and Grant Application Kit* released in December 2005.

The application used the DHS Target Capabilities as a framework for determining need. Each state and urban area was required to undertake a review of its capabilities with respect to national and, optionally, state/local priorities. The National priority capabilities are:

1. Information Sharing and Dissemination
2. Law Enforcement Investigation and Operations
3. Interoperable Communications
4. CBRNE Detection
5. Explosive Device Response Operations
6. WMD/Hazardous Materials Response and Decontamination
7. Mass Prophylaxis
8. Medical Surge

In addition, based on the NCR Strategic Plan, we identified the following six additional capabilities for our review.

9. Citizen Preparedness and Participation;
10. Citizen Protection: Evacuation and/or In-Place Protection;

---

[2] The 03 Urban Areas Security Initiative Parts I and II have been extended through November 30, 2005 and June 30, 2006, respectively.

11. Critical Infrastructure Protection;
12. Critical Resource Logistics and Distribution;
13. Mass Care (Sheltering, Feeding and Related Services); and
14. Planning.

Practitioners from across the region in each of the above areas completed capability reviews to identify strengths, weaknesses, and areas in need of improvement. From that process they identified resource needs and developed specific project concepts that would address those needs and strengthen the region's capabilities. From the $250 million worth of needs identified, we submitted an application for funding in the following areas. We reduced the amount requested from the fully identified need based on a vetting of the concept proposals and our capacity to execute.

Investment Area	Allocation
CBRNE Detection	$5.25M
Critical Infrastructure Protection	$26.25M
Citizen Preparedness and Participation	$10.00M
Citizen Protection	$11.00M
Critical Resource Logistics and Distribution	$21.00M
Explosive Device Response	$9.45M
Intelligence/Information Sharing/Dissemination	$10.05M
Interoperable Communications	$42.00M
Law Enforcement Investigation & Operations	$11.55M
Mass Care	$5.00M
Mass Prophylaxis	$3.67M
Medical Surge	$6.30M
Planning	$15.22M
WMD/Hazardous Materials Response & Decon	$11.50M
Total	**$188.24M**

HOMELAND SECURITY GRANT FUNDS SHOULD BE ALLOCATED BASED ON RISK AND EFFECTIVENESS

In FY 2006, DHS adopted a common risk and needs-based approach to allocating funds for the UASI program to strengthen national preparedness. We understand that DHS determined **risk** at the Federal level using a risk formula developed by DHS in conjunction with other Federal entities. The **need** was determined as an output of the Program and Capability review that we undertook for the region within the context of our strategic plan. Through the review process, the NCR developed two key submissions for the FY 2006 grant application:

1. *Program and Capabilities Enhancement Plan*, which is a multi-year global program management plan for the entire NCR homeland security program that looks beyond homeland security grant programs and funding; and

2. *Investment Justification*, which identified specific initiatives from the Enhancement Plan for which the NCR proposed to use FY 2006 UASI funding.

DHS used a nationwide peer review process to evaluate the effectiveness of each state or region's proposal in addressing the articulated need. We understand that the risk assessment accounted for two-thirds of the decision; the effectiveness assessment for the remaining third.

We support the direction DHS chose this year and agree that funds should be allocated based on these two criteria. However, the quantitative outcomes from the process suggest that the model did not work as intended.

A DHS shift to focus more on the risk of a state or region would suggest the NCR could only benefit from this change. Instead, this year, the NCR will receive 6.3% of the total funds allocated for the Urban Area Security Initiative program. Last year the NCR received 9.1% of the national total without the new emphasis on risk. The 6.3% the NCR is receiving this year is not the lowest percentage over the UASI's 4-year life span. The NCR received its lowest percentage of funds in FY04 when it was awarded 4.7% of the National total. That being said, the region as a whole expected to receive more than the 9.1% received in FY05 due to the new emphasis on risk.

We understand that DHS rated our proposal to be in the lower 50% for effectiveness. However, if the region was identified as a high risk region (in the 97th percentile of risk according to a DHS press release) but the effectiveness of the proposal was in question, it seems more consistent with securing the homeland that DHS would work with us to increase the effectiveness to address the risk rather than reducing our ability to address it.

NCR versus State Risk

The NCR was found to be in the top 25% of all eligible urban areas for risk. However, Maryland and Virginia as states were only in the top 50% for risk; the District of Columbia was in the bottom 25%. We can imagine that Maryland and Virginia fell lower in the risk category because they have both urban and rural areas. It is more surprising that the Department of Homeland Security finds the District of Columbia to be in the bottom 25% in terms of risk when the entire District, which houses the Capitol, the White House, the Supreme Court, and numerous national icons and critical functions, falls within the borders of the NCR, which was deemed to be in the top 25% for risk.

Peer Review Process

In terms of the outcome of the peer review process, the NCR prides itself on having some of the best trained, equipped, and experienced practitioners in the country. It is these practitioners who spent countless hours preparing and prioritizing the projects that fed into our application. For this reason it is difficult for us to understand how other practitioners across the country could find our application to be in the bottom 25% for effectiveness.

WE WILL BE UNABLE TO LOWER THE RISK TO THE REGION AS PLANNED

The impact of the FY06 funding decision from the Department of Homeland Security is that the region will not be able to accomplish as much as we had hoped in this funding cycle, which slows the momentum of our progress in building capabilities to better prepare the region. The following investments are examples of the broad capabilities we had hoped to advance fully in FY06:

- Mass Care
    - Following our experience in housing people who were driven from the Gulf Coast following Hurricane Katrina, we identified needs for enhanced coordination of mass care sheltering, including increased capability to maintain, deploy, and distribute critical supplies that would be required to support a significant sheltered population.

- Citizen Preparedness and Participation
    - Another lesson learned from Hurricane Katrina is that individual and family preparedness is an essential element of the region's preparedness and will impact our ability to respond and recover from a disaster. Our research shows that more than half of the region's population is not adequately prepared. We planned to invest in continued public engagement through training, media outreach, and supporting materials such as emergency checklists and supplies. We also planned to invest in strengthening of our volunteer corps, which leverages our capability to prepare citizens, and in improving our ability to ensure that special needs populations could evacuate or be supported in place as needed during as disaster.

- Critical Infrastructure Protection
    - Coordination of critical infrastructure protection is essential in our region since many of our assets cross through the three states. The Metro rail/bus system is a good example of a truly regional asset, and one for which we have an identified protection need. As a result of previous assessments, other transportation assets such as bridges and tunnels have identified protection needs for which we planned investment. We also had planned to expand our monitoring of water quality throughout the region to enable us to better and earlier identify natural or human-introduced agents that threaten public safety.

- Interoperable Communications
    - We had planned to invest more than $30 million to strengthen interoperable communications in the region. Included in that plan was $20 million to continue the establishment of secure, broadband connectivity and data exchange services throughout the region. This effort will allow real-time sharing of information throughout the region to allow police officers to access warrant information, transportation officials to monitor traffic flows during an evacuation, emergency medical providers and hospitals to track patients during a mass casualty event, emergency managers to work in a virtual regional emergency operations center,

etc. (The attached document further outlines our interoperable communications program and vision.)

We will not be able to support all of the investments above with the FY 06 UASI grant award from DHS, and there are many other investments in the areas listed on the previous page that we had planned as part of our $188 million application.

There is no question that many proposed projects within our priority investments will not be funded with FY06 grant money which will prevent us from increasing the security of the region as we had hoped. The capabilities we have built since 2001 will generally not be impacted. However, many of the enhanced, necessary capabilities we were hoping to build with FY06 money will be slowed or postponed. Below is a short list of the specific, crucial projects we proposed in our application. Because we were legitimately expecting an award of $100-110 million and only received $46.4 million, we will be unable to complete all of these projects in this grant cycle.

WMATA Alternative Operation Control Center ($9,000,000)

Currently, thirty percent of the region's commuters rely on Metrorail service, and almost half of peak period riders are federal employees. The recent attacks in London and Madrid have shown that transit systems are a favorite target of terrorists. WMATA's existing Operation Control Center (OCC) directs rail and bus operations, emergency repair actions, radio communications, coordinates communications with the region's emergency first responders, receives chemical sensor program data, origination point for public announcements, and needs to be extremely facile with the ability to quickly respond to a variety of incidents. WMATA's OCC represents a single point of vulnerability for operating the entire rail system. If the building that currently houses the OCC is destroyed or has to be evacuated, it would be essentially impossible to maintain rail service with any acceptable degree of reliability Addressing this single point failure in WMATA's operating system in a timely fashion will serve to mitigate the negative impacts and enhance the response and recovery capabilities of the National Capital Region resulting from a terrorist attack directed towards transit or other high threat targets in the region.

Metro Subway Security Strategic Initiative ($3,000,000)

Recently, the NCR Fire Chiefs Committee created the Subway Tunnel Working Group to identify gaps in Fire's response to an incident in the WMATA Metro System. Our current initiative being considered is to address identified gaps in the region's Fire Service abilities to respond effectively to an incident involving the WMATA tunnel system. This initiative includes implementing hazardous materials detection, mitigation and decontamination training to ensure proper CBRNE and hazardous materials techniques within the metro system are being met. This proposed project would also supply the first responders with longer duration breathing apparatus so firefighters can provide longer search and rescue missions within the tunnels in the case of an attack.

Increasing Special Needs Involvement in NCR Disaster Preparedness ($1,500,000)

The recent Katrina disaster in the Gulf indicated the need for disaster preparedness training for organizations like nursing homes, group homes, and providers of home health care that serve people with disabilities or other special needs. We recognize that these organizations in the NCR (or providing other services) are not sufficiently prepared to shelter in place the people they serve, or to evacuate them to a place of safety and shelter.

Our current initiative is to create a representative consortium of disability advocacy groups and service providers with national outreach and local NCR focus to prepare organizations that serve people with disabilities in NCR to shelter in place or evacuate those they serve and conduct exercises to shelter in place or evacuate to a distant shelter.

NCR Bomb Squad Training, Equipment and Caches ($8,000,000)

Currently, all NCR area bomb squads have a high level of interoperability and participate in information exchange, joint training, and mutual aid but lack adequate resources to obtain necessary operational equipment. As a result, all area state and local squads have immediate equipment deficiencies that impact the safety of the general public, emergency responders, and bomb technicians. Also, none of the state/local (S/L) squads in the NCR currently meet the minimum capabilities for Type I bomb squads as defined by FEMA.[3] This initiative, if funded, would allow all 8 area bomb squads within the NCR to obtain the equipment needed to meet the minimum standards of a Type I bomb squad as defined by FEMA.

Automated Fingerprint and Facial Recognition System (AFIS) ($3,000,000)

Mobile AFIS with facial recognition will allow NCR law enforcement officials in the field to identify persons of interest. With the use of the mobile device, a fingerprint and a picture will be captured and sent for search among the three participating databases consisting of approximately 1.5 million AFIS records. If the subject scanned has a prior local record, the result will be a "hit" on that record, returning the associated Mug Shot and name. This system is vital to protect the safety of law enforcement and first responders by allowing them to quickly identify dangerous individuals.

National Capital Region – Law Enforcement Information Exchange (NCR-LInX) ($3,000,000)

In June of 2005 the COG Police Technology Subcommittee collaborated with the Naval Criminal Investigative Service (NCIS) to launch the NCR – LInX after having explored multiple criminal justice data sharing options. The LInX system appears to be the most robust in the nation, having already successfully implemented criminal justice data sharing initiatives in the State of Washington, Hampton Roads Region of Virginia, Gulf Coast of Texas, the islands of

[3] FEMA Type I bomb squads are accredited bomb squads capable of handling multiple or simultaneous incidents, are CBRNE trained and equipped, and have 2 or more bomb response teams (BRTs). A BRT consists of 2 certified bomb technicians fully equipped for bomb squad operations. Teams must have render safe capabilities, including a robot capable of handling a Vehicle-borne IED render safe mission.

Hawaii and the area of the Florida – Georgia state line. The current initiative being considered would achieve complete criminal justice data sharing capability among the more than 75 Federal, State, and local law enforcement agencies that make up the National Capital Region. Without this system, routine criminal background checks across jurisdictional borders are delayed putting the safety of local law enforcement officers in jeopardy.

NCR Emergency Operation and Coordination Planning ($5,000,000)
This initiative will provide planning staff to assist the local governments in creating, updating and coordinating plans within the National Capital Region. Coordinating plans within the NCR is not only a priority of the region, but a priority of DHS and Congress. Both local and regional plans need to be updated in several specific areas in addition to a general review and update to make sure all information is current. Additional planning support will be primarily utilized to develop and or revise local evacuation plans, mass care/sheltering planning including animals centers, regional coordination and response plans, and continuity of government plans.

THE INDIVIDUAL JUSRISDICTIONS WERE ALSO IMPACTED BY THE FUNDING ALLOCATIONS

District of Columbia Impact

For the District of Columbia, we submitted a proposal for state funds in the amount of $21.8 million and were awarded $7.6 million. We worked collaboratively with the National Capital Region to develop our application to ensure we would leverage outcomes, gain greater sustainability, and be more cost effective. Overall District saw a decrease in funding of 46.7% from FY 05 while the program funding decreased by 29%. Also, even though the District of Columbia serves as the seat of the federal government and represents the core of the National Capital Region, the Department of Homeland Security found the District to be among the bottom 25% of all States and Territories in terms of risk. Only American Samoa, the Northern Mariana Islands, the Virgin Islands, and Guam received less funding than the District of Columbia through the State Homeland Security Grant Program.

We had planned significant investment to stand up our fusion center, which we have been developing in accordance with Department of Homeland Security guidance. Funds to support intelligence sharing and analysis, information infrastructure, and expansion of our terrorism tip program were included in that investment. We had planned investments in law enforcement to raise some of our critical capabilities to Tier 1 and to expand our rapid deployment forces. We had planned investment to develop our critical infrastructure protection program to enable strategic assessment of prioritized protection of the critical assets in the nation's capital. We had planned investment in expanded outreach and education to our residents, businesses, and visitors, to enhance their preparedness. And we had planned significant investment in continued planning, training, and exercising to ensure our first responders, support staff, leadership, and all relevant parties are ready to respond to the many possible scenarios that we face.

As a result of the announced grant award for the District, we will have to prioritize among the capability enhancements referenced above as well as the rest from our application, meaning that many will not get done within this grant cycle. Failure to execute all of the planned investments will not increase risk, as we will not lose those capabilities that we have already developed. However, we will not be able to continue building our critical capabilities and developing new ones as quickly as we had expected.

Maryland Impact

On May 31, 2006, the Department of Homeland Security (DHS) announced the allocation of $24.3 million in Homeland Security Grant Program (HSGP, which includes SHSP, LETPP, UASI, CCC, MMRS) funds to Maryland—a 38% reduction from last year's HSGP award of $39 million. Maryland requested a total of $120 million in its FY06 HSGP application ($69 million for SHSP and $51 million for the Baltimore UASI). According to DHS, Maryland's application was ranked in the top 50% for both risk and effectiveness. With that limited feedback, it is difficult to understand how Maryland's grant funds were cut so drastically.

Of the five grant programs that comprise the entire HSGP award, Maryland's biggest reduction was in its allocation for the State Homeland Security Program allocation (SHSP). In 2006, DHS allocated $8.1 million to Maryland—a 59% decrease which is greater than the overall nationwide reduction of 50%. One theory we have regarding Maryland's cuts is that the major risks in Maryland fall within the borders of the Baltimore UASI and the NCR UASI and that might have made the rest of Maryland fall lower in the risk category. We are looking forward to gaining more clarity on this issue at the After Action Conference which will be held by DHS in July, or sooner if possible.

Although Maryland's percentage reduction of SHSP dollars was less severe than only 13 other States and Puerto Rico, Maryland is not unique. Overall, DHS 2006 SHSP grants were reduced by slightly over 50%—including reduction of over 70% from 2005 levels for four jurisdictions.

Like 32 other grantees, Maryland is also faced with a reduction in funding for the Law Enforcement and Terrorism Prevention Program (LETPP). While allocating $384 million of 2006 LETPP grants (overall level funding from 2005), DHS reduced Maryland's portion of LETPP grants by $1.3 million—an 18% reduction from 2005.

Our Fusion Center (MCAC) will be our first priority for state and local portions of the grant. Some of Maryland's other priorities are Critical Infrastructure Protection, Agrosecurity, Interoperability, and Statewide planning efforts. Here are a few possible impacts, by investment, that are likely outcomes of Maryland's reduced funding:

- If Interoperability is not funded, it could impede Maryland's compliance with National Incident Management System (NIMS) by slowing the flow of critical information among first responders and emergency management agencies

- Without fully funding our critical infrastructure needs, the state will be unable to continue its implementation of the Interim National Infrastructure Protection Plan; maintain a dynamic regional collaboration program; or support the incentives and outreach necessary to maintain the development of the comprehensive information-sharing network between public and private sector security partners.

- Without the support of funding for intelligence and information sharing, Maryland will be limited on a state wide basis in its ability to share information/intelligence in a manner that allows for a timely and effective response to a threat. Without the support of this funding, there will be a distinct slower timeframe for the development and dissemination of intelligence and critical information. Additionally, Maryland will leave maritime assets which are vital to its homeland security vulnerable.

The program review processes designed and implemented by the Governor's Office of Homeland Security provides a suitable framework for the efficient review of the scalability of all local and state homeland security projects. Resources will have to be spread across the entire state and it will be difficult for Maryland to support all of its local jurisdictions while also enhancing capabilities in the Baltimore and NCR UASI regions. To be sure, the reduction in the 2006 award will cause delay in the completion of some projects. However the processes in place will enable Maryland to minimize those delays and direct the funds that are received to projects that provide the maximum overall benefits for the State.

Maryland's program development and review process allowed us to prepare for any amount of funds allocated by DHS. Although Maryland did not receive as much as was hoped for, the processes in place give us the ability to efficiently and accurately identify local and State projects which, when funded, will provide the greatest impact to the overall security and emergency preparedness of Maryland.

Virginia Impact

The Commonwealth of Virginia submitted investments for a total of $43.178 million and received $16.8 million. While this is substantially less than the previous year where we received $36.8 million (not $55.6 referenced in the invitation) rather than viewing this as simply a loss in funding, we consider it a reduction in the number and scope of projects that can be accomplished during this grant cycle.

Virginia's Homeland Security strategy has been developed through a partnership of the appropriate state agencies and the Secure Commonwealth Panel, composed of legislative, private sector, and local government representatives, to develop Virginia's overarching Homeland Security strategy. In January we convened a two day meeting of 250 subject matter experts from 50 professional public safety groups and private industry to develop specific initiatives and investments for this year's submission to the Department of Homeland Security. These were condensed into 13 investments, including five additional investments that were identified by the Commonwealth to be priorities. Inside the investments were 28 initiatives that identified current capabilities, goals for future capabilities, strategies for how to close the gaps, and funding requirements.

Despite these good efforts, the Commonwealth saw a decrease in funding of over 50% while the program funding decreased by 29%, even while the DHS risk-based portion of the formula placed Virginia in the top 50% for asset-based risk, top 25% for geographic risk, and top 50% for combined risk.

As a result of the amount of funding Virginia received, a number of our highest priority investments are in jeopardy of not being funded as mandatory requirements consume most of the grant allocation. This will not increase risk, as we will not lose those capabilities that have already been developed. However, it will take years longer to complete the critical investments that we identified and bring them to full operational status. As an example, the Commonwealth is in the process of completing its intelligence gathering and analysis Fusion Center, rated our highest priority for funding at $8 million, but we must now consider not funding through this grant cycle.

The Commonwealth of Virginia believed the new risk-based approach of DHS would be beneficial to the funding allocation we would receive as a result of Northern Virginia, Richmond, and Hampton Roads each being areas of concentrated risk, yet we scored only in the top 50% of all submissions. We followed the DHS guidelines and developed a comprehensive approach to the grant submission process, consulting with subject matter experts, convening large numbers of stakeholder groups, revising and consolidating our submission, and developing long-term strategies as well as short-term initiatives to address the gaps we identified.

We remain unclear on the DHS methodology and welcome the opportunity to learn more about the process so we will have a better understanding for future years. While we had expected more funding in this grant cycle, and have less ability to accomplish the critical tasks we identified, we will proceed as effectively and strategically as possible.

THE NCR WILL DISTRIBUTE FY06 FUNDS EFFECTIVELY AND STRATEGICALLY

We will use the $46.5 million granted from the Department of Homeland Security to continue strengthening the capabilities of the region to make it safer. While it will be more difficult to make decisions on funding, the money will be spent as effectively and strategically as possible. The state and local leadership have a conference call today to continue our joint discussions and we will be meeting in the upcoming weeks to make our final funding decisions. We also hope to meet with DHS to get a clear understanding of both the risk and effectiveness analyses of our application. We plan to participate in the DHS After Action Report Conference that will reveal the results of the peer review process All grant funds will be distributed by the end of August as required by DHS grant guidelines, assuming DHS has made the actual grants awards by the end of June.

Thank you for the opportunity to testify. As we know you are, we are fully committed and dedicated to a safer and more secure National Capital Region, and we work every day towards achieving that goal.

Chairman TOM DAVIS. Thank you very much.
Mr. Crouch.

STATEMENT OF ROBERT P. CROUCH, JR.

Mr. CROUCH. Thank you, Mr. Chairman, members of the committee. I do appreciate this opportunity to discuss this very critical issue, critical to the States of Virginia and Maryland and critical to the District of Columbia. Mr. Reiskin pointed out some of the investments that were included in our package and I think it might be useful to discuss some of the other proposals that are also included which we think demonstrate the forward thinking, seriousness and consideration made by the National Capital Region. We certainly know among al Qaeda's as targets in the past have been transit systems. We know that from the experience of London. We know that from the experience of Madrid.

Among our proposals is a Washington Metropolitan Area Transit Authority Alternative Operation Control Center. Currently, 30 percent of the region's commuters rely on Metro rail service, and almost half of the peak period riders are Federal employees. Recent attacks in London and Madrid have shown the transit systems are a favorite target of terrorists.

WMATA's existing Operation Control Center directs rail and bus operations, emergency repair actions, radio communications, coordinates communications with the region's emergency first responders, receives chemical sensor program data. It's an origination point for public announcements and needs to be extremely facile with the ability to quickly respond to a variety of incidents.

WMATA's Operation Control Center represents a single point of vulnerability for operating the entire rail system. If the building that currently houses the OCC is destroyed or has to be evacuated, it would be essentially impossible to maintain rail service within the acceptable degree of reliability.

Addressing this single point failure in WMATA's operating system in a timely fashion would serve to mitigate the negative impacts and enhance the response and recovery capabilities of the National Capital Region resulting from a terrorist attack directed toward transit or other high threat targets in the region.

Similarly, among our proposals is the Metro Subway Security Strategic Initiative. Recently the National Capital Region Fire Chiefs Committee created the Subway Tunnel Working group to identify gaps in their response to an incident in the WMATA Metro system.

Our current initiative is to address identified gaps in the region's fire services' abilities to respond effectively to an incident involving a WMATA tunnel system. This initiative includes implementing hazardous materials detection, mitigation and decontamination training to ensure proper hazardous materials techniques within the Metro system.

This proposed project would also supply the first responders with longer-duration breathing apparatus so firefighters can rely on longer search and rescue missions within the tunnels in the case of an attack.

Clearly one of the lessons of Hurricanes Katrina and Rita was the need to give additional attention to special needs populations.

Included among our proposals for the National Capital Region is a project to increase the capability of nursing homes, group homes and providers of home health care to service those people with disabilities or other special needs, including economic special needs. We recognize that these organizations in the National Capital Region are not sufficiently prepared to shelter in place the people they serve or to evacuate them to a place of safety and shelter.

Our current initiative would create a representative consortium of disability advocacy groups and service providers with national outreach and local National Capital Region focus to prepare organizations that serve people with disabilities in the National Capital Region to shelter in place or evacuate those they serve and conduct exercises to shelter in place or evacuate to a disaster shelter.

These are just three examples of the proposals in our package. We do think they demonstrate effectiveness. They demonstrate an attention to critical needs, unmet needs that require the attention of the National Capital Region and the Nation.

We are interested, as we know you are, in the process that was engaged in, in coming to the conclusions that ranked the National Capital Region as it has been ranked. Our colleagues, and we do regard them as colleagues, and partners at the Department of Homeland Security have assured us that we will have an opportunity in the very near future to have that kind of in-depth discussion.

As Deputy Mayor Reiskin indicated earlier, we are interested in the process, both in learning why our proposals were not ranked as highly as they were but also learning that in part so we can address those issues in coming applications. We want to get it right, and if we don't understand why we didn't seem to get it right this year, we want to learn that lesson.

Nevertheless, we do continue to have concerns about that process and whether the formula that is used is really the appropriate formula. In the Commonwealth of Virginia, we experienced a similar cut this year, dropping from over $36 million in funds last year to $16.8 for this year.

Again, the Commonwealth was ranked in the top 50 percent in risk, and we remain baffled why our proposals that included, for example, funding for our intelligence gathering and analysis fusion center, which is considered state-of-the-art among the Nation, funding for our interoperability programs. Virginia was the first State in the Nation to create a State interoperability coordinator, the first State to create a statewide interoperability strategic plan, which is being used today by the Department of Homeland Security as a model for other States.

We want to complete our interoperability throughout the Commonwealth, and yet these are proposals that did not merit a higher ranking and higher funding by the Department.

We continue to have concerns that the Hampton Roads region not only does not get Urban Areas Security Initiative funding but doesn't even qualify for consideration based on the formula, with all of its maritime and military traffic.

So we do continue to have those concerns. We look forward to that discussion with the Department and learning more about the process and contributing to its improvement in the future. Thank you very much.

Chairman TOM DAVIS. Thank you, Mr. Crouch.

Mr. Schrader.

STATEMENT OF DENNIS R. SCHRADER

Mr. SCHRADER. Mr. Chairman, thank you.

It is a pleasure to be here to give you our perspective in representing Governor Ehrlich's, his homeland security advisor, and appearing with my colleagues Bob Crouch and Ed Reiskin from the District and our local jurisdiction representatives who are here today.

Maryland is one of the three major components of the NCR, and one of the things we observed according to the Homeland Security application process is Maryland was ranged in the top 50 percent for both risk and effectiveness. One of the things that we are wondering as we sift through this is, it is difficult to understand how Maryland and the NCR grant funds were cut so drastically.

Our thinking is that we have to work on improving the process to ensure risk is properly calculated to realize a commonsense risk-based process, particularly since it's two-thirds of the application.

Federal Homeland Security distribution procedures changed this past year, and we support that. But the application seems to have gone into a black box of literally thousands of calculations and, in the end, fell short for Maryland and the NCR.

It appears that a laudable big picture goal which made a lot of sense in the front end of basing funding on risk instead got mired in complicated formulas, resulting in applications that don't match what we intuitively know to be high-risk areas.

The bottom line is more capabilities mean less risk. What we are really focused on is lowering risk, and by not building the capabilities, it slows our ability to reduce that risk.

Maryland, Virginia and the District jointly develop our security capabilities. We share information back and forth because it is a coordinated effort. We have three major airports, subways, ports, landmarks and, of course, the Nation's Capital.

Just to add on to what my colleague, Mr. Crouch, mentioned earlier in some of the projects that are on our list but will have to be scrutinized even more tightly than we already have, and we have done some pretty tight scrutiny; we spent an awful lot of time together, probably 20, 25 percent of my time with these folks in the National Capital Region because of the importance that the Governor places on this region.

Currently, all the bomb squads in the National Capital Region have a high level of interoperability. They participate in information exchange, do joint training and mutual aid, but they are not equipped up to the level that FEMA would expect for their highest level of standard, and we have an $8 million project which would address that issue in the National Capital Region.

There is an awful lot of pressure on our local jurisdictions. We partner with the local jurisdictions in the Capital Region, and they bear a lot of the workload in terms of helping us with the bottoms-up process of understanding what direction we need to go, and we need to make sure they have the planning resources so that we can coordinate our efforts because it is in effect coordinating the States and the District and the local jurisdictions in a very complicated coordination effort.

We have a program to continue to build our capabilities around evacuation planning, mass care and sheltering, animal shelters, regional coordination response plans, continuity of government plans, and we have $5 million projected for that.

We were disappointed in the State because primarily the three States—two States and the District—have to be strong to work together collaboratively. As I said earlier, we had expected somewhere in the neighborhood of $35 to $40 billion because we have a Central Maryland urban area as well as our law enforcement and State grant. We ended up with $24 million of which of course we are very grateful for, but it was a bit of a surprise, and we had put in an application for about $120, $69 for the State and $51 for the central Maryland urban area.

These are for key things like information sharing and intelligence. We have one of the Nation's first fusion centers that has been around since November 2003. Many of my colleagues from around the country have come to see our center and model it. Critical infrastructure protection, interoperability are just a couple of other major things that we need to invest in.

The bottom line is, at the State level, we have an open and transparent process just like we have here in the National Capital Region. We have the local jurisdictions help us to find the priorities, and we sift through that to—basically, the State has strategic objectives that we are focused on, and then we would hope the Department of Homeland Security would review those priorities and fund it as appropriate.

The bottom line is, we are looking forward to the July review session. We have reached out to DHS to start asking questions about these risk calculations. We do support this process, but we need to understand it better because the outcome in some cases just didn't make a lot of sense.

Thank you, Mr. Chairman.

Chairman TOM DAVIS. Thank you, very much.

Mr. Robertson.

STATEMENT OF DAVID J. ROBERTSON

Mr. ROBERTSON. Thank you, Mr. Chairman and members of the committee. I'm Dave Robertson, executive director of the Metropolitan Washington Council of Governments, and I have the privilege to represent the interests of 20 local governments in the National Capital Region that have vested countless hours and energy in addressing homeland security preparedness and response since September 11th.

Just yesterday, the Council of Governments' Board of Directors met to examine this issue and had a very thoughtful and helpful discussion. They, like this committee, have questions and concerns, and I believe they are seeking many of the same responses that the committee is seeking, which is, we need to know how to sustain what we have started and how it can be made more predictable and transparent for all those concerned.

I am here today, as I said, to represent the interests of our local governments but would comment to the committee that most response is going to be local. The men and women that are our first responders, our network of first responders in this region, work for local governments by and large, and they will continue to do what is necessary to protect and strengthen this region regardless of UASI funds.

We know we need these funds. We have identified a range of priorities, but in the event of an emergency, local officials will respond as they are required, no matter what the circumstances.

We will be looking to our State partners, our region's mutual aid agreements and certainly our Federal partners to support that effort. I can't speak too much to the peer review panel that worked on these applications, but I can speak to the peer team that helped reduce the region's proposals.

Tony Griffin, who is the county executive in Fairfax County, and his colleagues from around the region spent dozens of hours sifting through proposals and trying to make sure that they meet the region's top priorities and needs.

Local governments are supported by, among the Nation's best first responder teams: police officials, fire officials, transportation directors, health officials and others that worked hundreds of hours to develop these recommendations to measure them against standards and guidelines provided by the Federal Government. We know very strongly and very confidently that the work that has been advanced is of great need.

Speaking to the issue of predictability and sustainability, we are concerned that many of the projects that have been advanced may be jeopardized by the reduced funding. Two quick examples: Much of the equipment that has been purchased in the early rounds of UASI funding does sometimes have a shelf life or needs to be maintained. We will need ongoing funds to make sure that equipment is at the ready at all times.

Certainly, the issue of citizen communication. We can't do a one-time citizen communication campaign because this is a tremendously transient region. We need to reach out aggressively to citizens and the millions of visitors on an ongoing basis.

We also believe the impact on these UASI cuts will be serious for our region. Simply stated, this region cannot achieve the level of sustainability and predictability and confidence that we want to have in our region's preparedness with the funds that are out there. We have a tremendous unmet need that will be unmet, will not be recognized if we do not receive additional funds in our region.

Finally, the issue of the insufficiency perhaps of the application. Like a lot of the folks that have spoken earlier, the Council of Governments does support and our member governments do support the issue of risk-based assessment. We need to understand better how that is defined. We have worked very hard to address our proposals regionally and through the 37 capabilities that the Department of Homeland Security has identified. If there are other guidelines or additional information that will be coming forward in the next couple weeks and month, we look forward to working with State and local partners to pinpoint exactly how we can do better.

The Council of Governments was one of the first organizations to reach out to the Federal Government, State partners and locals immediately after September 11th to see how we could strengthen preparedness and response. We are proud of our work to date. We know our work is not completed; this is a long-term commitment of our Nation and region. And we look forward to working with Members of Congress to strength our preparedness.

[The prepared statement of Mr. Robertson follows:]

Testimony of
David J. Robertson, Executive Director
Metropolitan Washington Council of Governments

Committee on Government Reform
Congressman Tom Davis, Chair
Representative Henry Waxman, Ranking Minority Member

Regional Security: DHS Grants to the National Capital Area
June 15, 2006

Room 2154
Rayburn House Office Building
Washington, DC 20510
10:00 A.M.

Mr. Chairman, Mr. Ranking Member and members of the Committee: thank you for the opportunity to appear today to discuss the local impacts associated with reduction in homeland security funding. This is an issue of great importance to the governments of the National Capital Region (NCR), and we are grateful to the Committee for taking a closer look at homeland security funding decisions.

The Metropolitan Washington Council of Governments (COG) is a regional organization comprised of representatives of 20 local governments surrounding the nation's capital, plus area members of the Maryland and Virginia legislatures, the U.S. Senate, and the U.S. House of Representatives. Policies and plans are set by COG's full membership acting through the Board of Directors and its supporting bodies, as well as the National Capital Region Emergency Preparedness Council (EPC).

A major component of the COG vision is keeping the region safe and secure. Since the events of 9/11, the EPC has worked diligently to enhance our capabilities in emergency preparedness and planning. The residents of this region are now in a position of greater security, and we remain dedicated to effective, efficient and timely emergency planning to help keep them safe.

We are here today to assure members of Congress that:

- Local governments in the National Capital Region will respond quickly and effectively to emergencies in this area, using our own resources and in close coordination with state and federal officials.

- This region's leaders have invested homeland security funds wisely. We have purchased back-up sets of personal protective gear for first responders, we have adopted systems that allow those first responders to communicate with each other and regional leaders, we are working to secure and protect our water supply and we have emergency transportation plans.

- The Department of Homeland Security identified 37 capabilities that need to be addressed nationally, and DHS asked that the National Capital Region focus on eight of those priorities. Through our own rigorous planning process, officials decided to go above the call of duty by adding six additional target areas, making for a combined list of 14 capabilities.

- The impact of the cut in our UASI funds will be serious. Simply stated, we cannot achieve the security improvements this Congress values with the proposed funding level.

- Some have mentioned that the National Capital Region's application was somehow insufficient. Although we believe we followed the instructions we received, let us assure you that we will study all the feedback from DHS and that we are determined to strengthen the application. We are just as committed as we were after 9-11, when we were the first region in the country to respond with a detailed planning process.

Regional Security Risks

I would like to assure you that **there is no increased security risk to the region**. A major component of the COG vision is a firm commitment to keeping the region safe and secure. Since the events of 9/11, the EPC has worked diligently to enhance our capabilities in emergency preparedness and planning. Because of these efforts, the residents of this region are now in a position of greater security.

We remain dedicated to effective, efficient, and timely emergency planning which can help keep residents safe, and we know that local governments in the National Capital Region will respond quickly and effectively if there is an emergency in the area. We have worked diligently over the past several years to ensure that adequate plans are in place. These homeland security plans, used in collaboration with state and federal officials, are more than adequate to handle a potential crisis situation.

Investment of Homeland Security Funds

The Urban Areas Security Initiative (UASI) grant program of the Department of Homeland Security (DHS) provides financial assistance to address the unique planning, equipment, training, and exercise needs of high risk urban areas, and to assist them in building an enhanced and sustainable capacity to prevent, respond to, and recover from threats or acts of terrorism.

Regional leaders have **worked collaboratively to wisely invest homeland security funds**. Our primary focus, once again, has been to put in place effective programs and policies that will keep the region safe for years to come. We have purchased back-up sets of personal protective gear for first responders and adopted systems that allow those first responders to communicate with each other quickly and directly. We have developed emergency transportation plans, and are working to secure and protect our water supply.

The table below shows how UASI funds have been allocated to the NCR since fiscal year 2003. As you can see, the proportion of funds deemed by DHS as necessary for NCR preparedness efforts is significantly lower than what was granted to the region in fiscal year 2005. As such, the proposed funding level is much lower than what NCR officials had deemed would be necessary to efficiently build upon the solid groundwork that has been laid for the safety of the region.

Fiscal Year	NCR Allocation	Percentage of Total
FY 2003	$60.5M	10.3%
FY 2004	$29.1M	4.3%
FY 2005	$77.5M	9.1%
FY 2006	$46.5M	6.3%
Total	$213.8M	7.9%

Impacts of FY06 Funding

The table below, which outlines the funding requests that were included in our application to DHS, shows that **the region still faces a tremendous amount of unmet needs.** The officials who set these priorities are among the most the experienced in the country, and it is their belief that the region **stands to be impacted by the decreased level of funds.** In addition to training and equipment for our first responders, the NCR had hoped to dedicate a large portion of funds to new technology that would connect these first responders in the event of a terrorist attacks or other emergency. While reduced funding will not make the region less safe or secure, it will limit how quickly we are able to build upon the solid foundation we now have for connecting our law enforcement officials, community leaders and other stakeholders.

Investment Area	Allocation
CBRNE Detection	$5.25M
Critical Infrastructure Protection	$26.25M
Citizen Preparedness and Participation	$10.00M
Citizen Protection	$11.00M
Critical Resource Logistics and Distribution	$21.00M
Explosive Device Response	$9.45M
Intelligence/Information Sharing/Dissemination	$10.05M
Interoperable Communications	$42.00M
Law Enforcement Investigation & Operations	$11.55M
Mass Care	$5.00M
Mass Prophylaxis	$3.67M
Medical Surge	$6.30M
Planning	$15.22M
WMD/Hazardous Materials Response & Decon	$11.50M
Total	**$188.24M**

Application Evaluation

We generally support the risk-based approach taken by DHS, but believe the Department's risk and effectiveness-based methodology used in determining fiscal year 2006 grants to be flawed. In assessing applications, DHS first looked at the level of risk in particular areas -- which accounted for two-thirds of each final evaluation. For the remaining one-third of the evaluation, the Department then considered the effectiveness of applications in terms of their relevance, innovation and feasibility.

Using this formula to allocate funds to the NCR, the department determined that each individual element of the application -- both risk and effectiveness elements -- was either at or above average. Yet, somehow, the Department's compounding of these above-average elements resulted in a final determination that the effectiveness of the overall NCR application was "in the bottom 25 percent of all Urban Area submissions," even though the region's risk was found to be in the top 97th percentile of all eligible areas. The computation of these elements, as well as some of the short shrift of the review, is a process the officials of the NCR cannot understand.

The region's homeland security plan is complex and constantly evolving, and we are committed to studying all feedback from the Department of Homeland Security to make the plan as effective as possible. Although we believe we followed the instructions we received, **we are determined to strengthen the application**.

Distributing Funds

In closing, I would like to reemphasize that the capabilities we've built over the last four years will not go away. The region will not be less safe or secure as a result of lower funding levels, and the EPC will remain as committed as ever to keeping the region safe. Even though, with this level of funding, we will have less ability to accomplish as much as we expected to further enhance the region's preparedness, we will proceed as effectively and strategically as possible.

With this in mind, the Senior Policy Group (SPG) and Chief Administrative Officers (CAO) will prioritize needs to develop a list of the projects that can be funded by the current award level of $46.4 million. Although it will be a challenge to determine how to best distribute funding, stakeholders will do everything in their power to ensure the money is spent as effectively and strategically as possible. While it is our belief that the highly the experienced practitioners who worked on the application did follow guidelines provided by DHS, we will examine all feedback and work to strengthen the document.

Chairman TOM DAVIS. I want to thank all of you for your testimony.

Let me start the questioning, Mr. Foresman, with you. DHS rated the National Capital Region in the top 25 percent for risk but, as has already been stated, the District of Columbia was rated in the bottom 25 percent. Considering that the District has already been a terrorist target, it houses the Federal Government, the White House, the Nation's Capital, the Capitol building, Supreme Court, why was the risk rating for D.C. in the bottom 25 percent?

Mr. FORESMAN. Mr. Chairman, thank you for the question. This actually goes to the importance of us making sure that we are very clear communicating to our State and local partners. You are actually referring to two separate grant programs and the analysis contained therein.

In terms of the rating for the urban area, D.C., which is based predominantly on the risk structure, the District of Columbia and the National Capital Region was rated in the 97th percentile.

D.C. also is unique as a city; it has both the State status and city status, and so it is eligible to participate in the State Homeland Security Grant Program. So that program has a different funding formula that is based, in part, based on a floor amount, population-based floor amount.

So, in essence, D.C., from a ranking standpoint, is competing against 50 States from a population base before we even get out on the door on any of the risk discussions. So it is an artificiality of the fact that D.C. is lumped into what is a State-directed program with a population-based funding formula.

Chairman TOM DAVIS. So the risk is there. I mean, a major criteria of risk under the State funding formula is population.

Mr. FORESMAN. A major criteria, that is correct. There is also a floor funding for the State program. There is a floor funding before you can get into the risk discussion.

Chairman TOM DAVIS. So you really think under a State formula that Montana is at higher risk than D.C.?

Mr. FORESMAN. Congressman, there are all sorts of intricacies.

Chairman TOM DAVIS. I just asked you a question.

Mr. FORESMAN. Yes, sir. My answer is, we do not measure the risk of one community against the other and place the importance higher—the higher of one community against another.

Chairman TOM DAVIS. Am I right, under the State program, you have two criteria, and one is risk, and on the risk side, I can understand looking at the grants, I'm not going to get, at least in this question, how good the city's grant was, but on risk analysis alone, doesn't it seem on risk analysis alone that the District of Columbia is a likelier target and at higher risk than the State of Montana, or North Dakota, where I'm from, if a terrorist can find North Dakota?

Mr. FORESMAN. I understand. The point is a valid point. If it was purely risk, D.C. would have scored higher. But there is a population-based factor on the front end of it, that amount, for a large portion.

Chairman TOM DAVIS. You call it risk analysis. I mean, what it says on the grant, it says risk; it doesn't say risk and population. There is risk, and there is effectiveness.

Mr. FORESMAN. Mr. Chairman, I go back to what I said earlier in terms of the products we provided to the communities. We should have been clearer in our communication, and that does not effectively articulate the nature of the grant program and the allocation process.

Chairman TOM DAVIS. So, at a minimum, you would agree with me that where it says risk, it shouldn't say risk because, otherwise, if you're saying risk and population is a factor, that's one thing, but you have ranked the District under the State categories as a lower risk than North Dakota.

Mr. FORESMAN. Mr. Chairman, we could have provided a better and clearer descriptive response to the community.

Chairman TOM DAVIS. Risk is really not a very good term there.

Mr. FORESMAN. Mr. Chairman, this goes to the broader issue. Throughout this whole debate, a number of people have been confusing threat and risk in the same discussion, and risk analysis is not risk management. Threat is not risk. What is important for us——

Chairman TOM DAVIS. What is risk?

Mr. FORESMAN. Risk is the combination of threat, vulnerability and consequences. It is the threat, the intention of the adversary, the likely intention of the adversary to commit an act, the consequences, what's the likely outcome, and the vulnerability is the relative measure.

Chairman TOM DAVIS. Under that criteria—I say that's a fair definition of risk. Let's agree that's fair. The city's population is concentrated, so one incident there is going to hit a lot more people than some of these western States where people are all over the place. It's a much more likely target.

How in the world is, a risk or a threat in Montana, North Dakota or South Dakota higher than the District? I just need to understand that. They have a few more people there, but in terms of any kind of risk analysis, I don't think that passes the laugh test.

Mr. FORESMAN. Congressman, this comes down to the fact that D.C. falls into two categories. It falls into a State category in some of these grant programs. It falls into an urban areas category in some of these grant programs. Deputy Mayor Reiskin and I have talked about this over the course of the last several years, and it creates a little bit of an artificial challenge.

Chairman TOM DAVIS. I understand that DHS made adjustments—how much was in the State versus how much was in the urban category in terms of the amounts that DHS was throwing in? You have two different grant categories; how much went to the urban, and how much went to the State?

Mr. FORESMAN. In terms of total dollars?

Chairman TOM DAVIS. Yes. Or percentage wise.

Mr. FORESMAN. The urban area was about a little over $700 million and the State program was about, I believe, $500 million.

Chairman TOM DAVIS. All right. I understand that DHS did make adjustments to the risk factor assigned to the National Capital Region and New York City, based on a DHS briefing of our staff. Is that correct? That after the adjustments were made under the formulas, it looked like New York City and the District were lower, and you made what we used to call Kentucky windage adjustments.

Mr. FORESMAN. This goes back to the discussion on common sense. Because we understand that we are responsible, and Congress's direction is to protect the Nation's risk. When we take a limited pot of money and we have to apply that against an entire risk of a Nation, we wanted to ensure that we did not reduce the National Capital Region funding and that of New York City below the average of what they received in past years because, frankly, this is the first year that we really quantitatively had a good understanding of risk outside of the National Capital Region in Washington, and that changed how we had to allocate the money accordingly.

Chairman TOM DAVIS. Let's just take a look at the city side where you say the city is treated fairly here; that is the thesis here, isn't it? That the National Capital Region under the urban formula is going to get treated fairly? Even under that, as I understand it, Washington, DC, wasn't first, wasn't second, wasn't third, wasn't fourth. I was ranked below New Jersey; it was below the Bay area; it was below Los Angeles, and so on. And in fact, you included additional urban areas like Ohmaha that ate up money that could have come here.

Mr. FORESMAN. Let me use an example without specific jurisdictions. There's a large metropolitan area that has roughly the same number of assets in terms of critical infrastructure as the National Capital Region, and it has a population that is roughly the same.

The National Capital Region population, from a density standpoint, is spread over 3,000 miles, and in this major metropolitan city, it's spread over 230 square miles. So the net effect of an event in that particular community would have been potentially far more grave in terms of an immediate impact on a large concentrated population. This goes back to the whole discussion about New York City with its high population density.

In that particular case, as we ran the numbers, when you looked at the risk ranking, it raised that city above the National Capital Region. But there's an intrinsic value of the Nation's Capital psychologically to the American value system, and that is the key reason and one of the key reasons that we put, as you say, the Kentucky windage to it, because risk management, risk analysis, it is not absolute. It only seeks to guide and inform our decisionmaking process and not control or dictate it. So we were trying to——

Chairman TOM DAVIS. I appreciate your explanation, but I am going to yield to Ms. Norton in a second.

I just tell you, from my perspective, you are looking at the Nation's Capital, which has been hit once, had the anthrax attack on top of that. We are still I think on everybody's—close to the top of everybody's—hit list for any terrorist group that wants to make a statement.

Terrorists are hitting London and Madrid. They are hitting capitals. They're not likely to hit Bull Frog Corner, WV.

The reality is, as you take a look at urban areas and our density and everything else, I think your bean counters are just counting the wrong beans. I mean, I just think that this is too formula driven. I do not think that Congress makes the right decisions when we start allocating money and earmarks; it's not done on an equitable basis. That is why in this particular area, when it came to allocating funds for Homeland Security, we tried to give it to the administration because we thought they would be fairer, that they would take a look at what was good for the country as a whole, take a look at the threats and try to take the politics out of it. But what you have replaced with the politics is some bean-counting formula that doesn't pass the laugh test.

You have, under your procedures, Montana, North Dakota, and South Dakota ranking ahead of Washington, DC, and under the regional analysis now, the city is not even second—now they are fifth—and that is applying, as we say, some subjective factors. But more importantly, you are including other urban areas that could have been covered under State grants that frankly don't look like they are threatened at all.

What I'm concerned about is, should an incident, God forbid, happen—and we hope it doesn't—if it happens in Omaha, I think you're covered. I think you can say, we took care of them. But should something happen in New York or Washington, I think you are going to hear a huge outcry that we didn't give the appropriate attention. And the problem is, in these areas, they've been hit before; they've been hit more than once. That is the likelihood, and that's our concern.

Ms. Norton.

Ms. NORTON. Thank you, Mr. Chairman.

Picking up on your questions, by the way, I certainly hope al Qaeda doesn't notice these rankings and the shift. They're informed more than Members of Congress.

The chairman asked questions about the District. I'm curious, did you, in calculating risk for the District of Columbia in particular, take account of the fact that the District's daytime population is two or three times its census population?

Mr. FORESMAN. Congresswoman, we do take in the computing population in the day and night population, and again, it depends on which of the programs.

Ms. NORTON. I'm talking about the area grant.

Mr. FORESMAN. The urban area, yes, ma'am.

Ms. NORTON. So then, of course, we're talking about between 15 and 20 million people a year who come; we're talking about a population that is not 600,000 but three times that. That's the population figure you used?

Mr. FORESMAN. Congresswoman, I don't have the exact figures in front of me, but when I was serving in Virginia, whatever the current population of the National Capital Region is, with some slight increase for tourist population, is going to be the number that you see used, and I don't whether it's 4 million or 5 million.

Ms. NORTON. I was asking you about the District of Columbia.

Mr. FORESMAN. But the District of Columbia, Congresswoman, this is a very important distinction because the District of Columbia under the Urban Area Security Initiative program is part of the National Capital Region, and it looks at the combined total of all of the jurisdictions.

Ms. NORTON. If you look at the District of Columbia grant itself, did the District of Columbia grant take into account that what you're protecting is the daytime population, in effect?

Mr. FORESMAN. Congresswoman, I don't know specifically on the State Homeland Security grant program, but I will find out which of the two numbers was used in the formula and give you a written answer.

Ms. NORTON. Very much appreciate that.

Mr. Foresman, I know you would always, when grants are being made, expect some disagreement. I believe there would have been less disagreement or certainly less outrage if there had perhaps been cuts but they had been more modest.

The impression created is that we got cut Homeland Security; who got the money, let's cut them. So without some explanation, it looks like you said, where's the money, and the money is in New York and the National Capital Region.

This is my question. I have not heard you indicating in your formula what I will call, for lack of a better term, the price of postponement. In a real sense, if we know what we all know about the region, what al Qaeda and the terrorists know about this region, my question is: Isn't there some reason to make haste in places which are particularly at risk to get it done, as it were, to get its most vulnerable parts staunch down as opposed to looking at, of course, the various vulnerabilities across the country? Some have not been as adequately funded as you would like.

But if you factor in what a single incident would mean here because we hadn't gotten to it as opposed to making sure that you take care of what is vulnerable in other regions, isn't there a value that the Department would put on getting as much of it done as rapidly as possible in this region and perhaps in New York, the other target that everyone is aware of?

Mr. FORESMAN. Congresswoman, let me address it this way; that is a large part of why you see a major change. If you look at the five high urban areas: Los Angeles/Long Beach region, Chicago, the Newark region, San Francisco, the combined increase because of the understanding of the risk score in those communities, very densely populated areas with very critical infrastructure, that the change in allocation based on the risk scoring alone was $38 million.

So part of what I offer to you, Congresswoman, is while everybody talks about Omaha or Louisville, they represent a very low dollar amount in the context of how these dollars were adjusted. A large amount of dollars went to these high metropolitan areas, recognizing that Chicago, for instance, is a major financial and transportation center, a major manufacturing center.

So our concern is about making sure that we bring percent readiness up, and that we don't have any weak links or Achilles' heels, particularly in those urban areas.

Ms. NORTON. I looked at the other cities, New York, LA, Chicago. I don't understand Florida. You will have to make me understand that one. Then we get to D.C. I said, well, population, I can understand that.

I have to tell you that with all of its population in LA spread out the way it is, they are now only making a downtown in LA, one only wonders if MO of al Qaeda is simply to look for population or whether it's not to look for concentration of icons and population.

Why New York in the first place? Why was New York hit twice? Why was the same building hit twice?

Mr. FORESMAN. Congresswoman, that goes to a key point. One of the great criticisms after September 11th was no one thought about a plane being flown into a building. We have heard bin Laden say on numerous occasions since September 11th he wants to hurt America physically, psychologically and, most importantly, economically.

Take the LA area as a good example. It is spread out like the National Capital Region but with a higher density. It is a major center of economic activities.

Ms. NORTON. I have to stop you right there for a minute. They may well have gotten cut. What I indicated was, I'm asking you if you recognize that al Qaeda does have an MO, a known MO. They are looking for large concentrations of people that you can hit. We are not talking about the atom bomb here, so you can't just hit LA and expect it—we're talking about somebody that wants to hit someplace, and by hitting that one place you get an icon or you get huge numbers of people.

The reason there is more chatter about New York than D.C. is because you hit a few blocks there, you have so many folks that you don't need to worry about LA, San Francisco or even Chicago.

So I'm asking, with this notion of population, whether or not you are really focusing in on density, where al Qaeda has been focused, not just population.

Mr. FORESMAN. We are focused on population. We're focused on population density. But, Congresswoman, as a good example, the day before Hurricane Katrina, one of our major metropolitan areas in America constituted 25 percent of the chemical production capability in America. The day after that, it constituted 45 percent because—not chemical, but petroleum—because of the number of refineries that had been knocked out of service in Louisiana.

What this points out to us is that there is a physical impact on people, but we also have to be concerned about the impacts on infrastructure, because you and I both felt the pain at the pump right after Hurricane Katrina when we had to go in and pay more for gas because the petroleum industry's capabilities had been reduced.

Ms. NORTON. Did you take into account, because you just named infrastructure, Madrid and London and the vulnerability of WMATA and its tunnels when these cuts were made?

Mr. FORESMAN. We have taken into account the critical infrastructure here, and if you look at the other metropolitan areas where you have subway and metro systems, some that are subject to worse flooding than metro, as a prime example, we take all of these factors into consideration.

Chairman TOM DAVIS. Thank you.

Mr. Gutknecht.

Mr. GUTKNECHT. I yield to my colleague from New York.

Chairman TOM DAVIS. Mr. Fossella.

Mr. FOSSELLA. Thank you, Mr. Chairman. Thank you for the indulgence here.

We are here, as you all know, because of what happened on September 11th. The Department of Homeland Security exists because of what happened on September 11th. It happened in New York and Washington, DC, and obviously in Pennsylvania.

There are many who I think share the goal that the fundamental responsibility of the National Government is to protect innocent people. That was the intent of the Department of Homeland Security. We are here today in addition to answer the question fundamentally how we have gone astray.

By way of background, I represent Staten Island and Brooklyn. More than 200 people were killed on September 11th; 78 of the 343 firemen were from Staten Island; 29 people in my parish were killed. And I just don't want to have another day of mourning, another September 11th.

I have been wholehearted in support to ensure that not one family in America ever again has to fall victim to an act of terror. So when we wake up and we realize that New York City sees a reduction percentage wise of upwards of 40 percent while other cities around the country, arguably not as high on the threat list, receive increases of 20, 30, 40 percent, we have to ask the fundamental question, are we doing what's right?

The original intent of UASI went from seven cities to 35 to 46 regions that now cover over 100 cities; is that right?

Mr. FORESMAN. That's correct, Congressman.

Mr. FOSSELLA. Are we not drifting from the original intent of much of what USAI was meant to serve, that is, the areas like Washington, DC, like Chicago? I'm not saying here that New York deserves 100 percent, or Washington, DC, deserves 100 percent. Just follow the threat, follow the risk.

Mr. FORESMAN. Congressman, I think it's a very reasonable question that you have asked here. The 46 urban areas represent about 85 percent of the Nation's urban area risk, and why this is important—and part of this gets into the debate that you go for 80 percent, 90 percent, 85 percent. Where we saw the risk measures have a measurable decline to where there was a clear separation was about the 85 percent point, and that was those 46 urban areas.

I don't think there was any intent of the program that I'm aware of back when it was first established to say it was going to be 7 versus 50 versus 46 versus 30, but I think that we have to get beyond the discussion of funding on a single year and look at it over the course of funding activity and look at it in the context of managing risk. And we are responsible for finding a way to manage America's risk, and the way we best feel we can do that is to apply those dollars to get the maximum reduction in our nationwide risk, and that's how we came up with the 46 urban areas.

Mr. FOSSELLA. Having said that, in all the intelligence chatter, I'm not asking you to divulge anything classified, is there any city that is more targeted than New York City? Not just in this country but in the entire world.

Mr. FORESMAN. Congressman, I don't know the specific answer to that. I will make the assumption that probably not, but I won't make it exclusively, but we should not equate threats with the complete discussion on risk. Threats are one component to it, but we have clear indications of al Qaeda and other groups' intents;

through the radicalization and the globalization of the al Qaeda activities, we have clear understandings of their intent to wreak economic, physical and psychological pain elsewhere.

Mr. FOSSELLA. But am I asking too much to indicate that New York City is still the No. 1 threat?

Mr. FORESMAN. It was the No. 1 risk ranker last year; it remains this year.

Mr. FOSSELLA. So if you sit back and know, far and above, with the two attacks in 1993 and 2001, that New York City is indeed No. 1, isn't there anybody sitting back saying, well, why are they seeing a reduction in funding? I'm just curious, nothing about the good people of Omaha, but are they in the top 10 or 20?

Mr. FORESMAN. Congressman, I think one of the things I would go to is this particular pie chart. If you look at New York City alone in terms of the total allocation of the UASI program from fiscal year 2003 through fiscal year 2006, you know they have not quite gotten a quarter but almost a quarter of the dollars have gone to New York City alone, and they have about 14 percent of the Nation's urban area population.

But what I will tell you is, we are putting our heaviest focus and our heaviest attention in terms of financial commitments into New York City. We have and we can continue to do so this year, and it's averaged twice as much as other—as the next highest jurisdiction, not quite——

Mr. FOSSELLA. The reason is because that's where the risk and the threat is. Whether you live in New York or have family who visits, the millions of people from across the country, if you have a daughter or son who's a tourist, I think you would want them to be there.

I have one other question.

Mr. GUTKNECHT [presiding]. I have to intervene here because we are going to try to keep the hearing going through the vote. We have a vote on the floor of the House, and the chairman has run over to vote, he's going to come back, and we're going to go. I really need to let Mrs. Maloney have her chance. We can have a second round. If you want to run and vote, you can come back, and I think the chairman would be more than willing to entertain questions.

Mrs. Maloney.

Mrs. MALONEY. Thank you, and I support my colleague's line of questioning. He has joined, as I have, the chairman of Homeland Security Peter King in asking for a meeting with Secretary Chertoff. We sent that letter off in a bipartisan effort roughly 2 weeks ago. We have not heard back. Can you give us any indication when he'll be able to meet with us?

Mr. FORESMAN. I don't know the specific date the meeting has been scheduled, but it's my understanding when I was last briefed on that a couple of days ago that the Secretary was more than willing to meet with the delegation, and it may in fact be early next week.

Mrs. MALONEY. I would like to go over my understanding of this year's process for distributing the Homeland Security funding. Let me see if I have this correct.

Cities and States prepared an application for submission to the Department of Homeland Security, making their best case as to their threat, need and vulnerability. However, the individuals preparing the applications frequently do not have security clearances and, therefore, a lack of access to real-time information about true risk and vulnerability their city or State may actually face. It is therefore the job of the applicant to give their best guess on what their need is, absent a classified threat assessment.

After the applications are submitted, they are analyzed by a peer review board, and it is the job of the peer review board to analyze and rank applications on the basis of risk, vulnerability, among other things.

Once again, due to the sensitive nature of some of the information involved and the lack of security clearances among members of the peer review board, they also lack real-time information regarding the actual risk or vulnerabilities to the cities and States being reviewed and ranking.

I have been informed that the methodology for the peer review process was developed and implemented by a consultant firm hired by the Department of Homeland Security, reportedly Booz Allen Hamilton, and it is my understanding that it was this consulting firm that administered the review, collected the review, scored the sheets following the review and ranked State and urban area applications.

I have also been informed that it was only after Booz Allen Hamilton completed their analysis that the Department of Homeland Security became involved in the process by signing off on the analysis and allocated funding based on Booz Allen Hamilton's works.

So my first question, Mr. Foresman, is my understanding of the process correct?

Mr. FORESMAN. It is not very correct at all, Congresswoman.

Mrs. MALONEY. Let me ask you, do State and city application writers have access to the threat information to best prepare their applications?

Mr. FORESMAN. Congresswoman, if you might, I can maybe try an explanation to capture all of it.

Mrs. MALONEY. If I could, answer——

Mr. FORESMAN. There are two very distinct points. The application writers do not have the specific threat analysis data or the risk analysis data in front of them, but it's not necessary in the context of the preparation of the application.

Mrs. MALONEY. Let me just ask you, do the members of the peer review process have security clearance and access to real-time threat and risk assessment?

Did they have security clearance and access to real-time threat and risk assessment?

Mr. FORESMAN. Congresswoman, I won't give you a blanket answer, but I'll be more than happy to give you a written response on whether all, some, none. But what they were looking at was not the risk piece of it. Let me be very clear. In terms——

Mrs. MALONEY. Why in the world were they not looking at the risk piece of it? Your application says risk analysis, risk-based funding.

Mr. FORESMAN. Congresswoman, there are two parts to the analytical process. The first part is the risk analysis. The risk analysis is conducted by the Federal interagency, if you will, predominantly by the Department of Homeland Security. That is based on the classified information, the unclassified, the open-source information, the data sets, the population data. That forms one piece of the allocation equation.

On the effectiveness score, the peer reviewers were provided, in terms of the application packages, all the applicants were given a package that says, here are the national priorities, here are the things by which your application will be rated toward those national priorities; you understand your risk, what you need to do locally. And we've asked them to submit the peer review data.

Mrs. MALONEY. My time is running out, and I would just like some questions asked, specific questions. Do members of the peer review process have access to and sign off on the final rankings of the applications, yes or no?

Mr. FORESMAN. No.

Mrs. MALONEY. Did the Department of Homeland Security have a contract with Booz Allen Hamilton to conduct the peer review of these grants, yes or no?

Mr. FORESMAN. Congresswoman, that cannot be answered with a yes-or-no question.

Mrs. MALONEY. Did Booz Allen have a contract with the Homeland Security Department?

Mr. FORESMAN. Booz Allen had a contract with the Department of Homeland Security to facilitate the peer review process, not to do the evaluation of the investment justification.

Mrs. MALONEY. Can you provide me and the committee with a copy of that contract?

Mr. FORESMAN. We'd be more than happy to, Congresswoman.

Mrs. MALONEY. Was the Department of Homeland Security an active participant in the peer review process, or was that responsibility contracted out to Booz Allen Hamilton?

Mr. FORESMAN. We used Booz Allen Hamilton to help facilitate the peer review process, but there was a second level of review by Federal Department of Homeland Security employees of the scoring, and they were also present during the scoring processes.

Mrs. MALONEY. Who actually ranked the States and the urban areas? Was it the peer review process, Booz Allen or the Department of Homeland Security?

Mr. FORESMAN. The Department of Homeland Security, based on an amalgamation of information from the risk analysis and the peer review process.

Mrs. MALONEY. Who in the Department of Homeland Security made that decision?

Mr. FORESMAN. Congresswoman, I don't know that it was any one person, but I will get you the specifics of who was involved in that.

Mrs. MALONEY. It appears that you shifted the formula from risk to need. Because if you based it on intelligence and real risk, there's no question that it would be New York City and D.C. I mean, did you consult the list of targets from al Qaeda and the intelligence from al Qaeda that has been reported in the papers? And some of us have seen it where all of the chatter is primarily, three-fourths, New York. Then maybe they get down to D.C. But by all security accounts, whether it's the FBI, the CIA, the National Intelligence Director, the 9/11 Commission, all of them report that all of the intelligence, that the risk is New York City. The threat they believe is a bomb that would be put off that would kill hundreds of thousands of people.

I lost—as my colleague Mr. Fossella and we've worked on many responses to September 11th together—I lost 500, he lost 300. All of us lost people in New York. We appreciate the help of our colleagues here in Congress, but this formula does not reflect the reality of what's out there or the reality of risk.

I'm all in favor of a State aid formula to help States, but don't call it Homeland Security, call it municipal overburden. Call it aid to States, but don't call it Homeland Security.

I look forward to our meeting with Mr. Chertoff, and my time is up.

Mr. FORESMAN. Thank you, Congresswoman.

Chairman TOM DAVIS [presiding]. I have a few more questions. DHS said it worked closely with the urban areas applying for funding. It also provided a guide outlining the application process.

Mr. Reiskin, what role did DHS's office of the National Capital Region play in assisting you with the application process?

Mr. REISKIN. The application was developed, first of all, being guided by a strategic planning process that was in part guided by the Office of National Capital Region Coordination.

Chairman TOM DAVIS. DHS helped you put together the plan, basically.

Mr. REISKIN. They were an integral part. That office was an integral part, really leading the effort to develop the strategic plan guiding the region for the next 3 to 5 years. That was the starting point for the process.

From there, the senior policy group which includes the Office of National Capital Region Coordination and the chief administrative officers, so, basically, the State and local leaders outlined the direction for the application, identified the target capabilities around which we would build the application. There were eight national priorities. We added six regional priorities, and then we let our subject matter experts, the police chiefs, fire chiefs, transportation directors, do the analysis of the region's position in each of those capability areas.

That was all brought back together and again reviewed by the State and local level leaders, which includes the Office of National Capital Region Coordination.

So throughout the process, from start to finish, they were a part of the process.

Chairman TOM DAVIS. Did they give you any suggestions that you didn't take in putting the application together?

Mr. REISKIN. I would say that the office was not acting as really an agent of the Office of Grants and Training; they were acting really as our partner, our Federal partner in the region, not really serving in the role of advising us on the grant process itself.

Chairman TOM DAVIS. But the plan to which the grant applied was coordinated with DHS, correct? The regional plan?

Mr. REISKIN. The regional strategic plan was coordinated with the Office of National Capital Region Coordination.

Chairman TOM DAVIS. Which is part of DHS.

Mr. REISKIN. Part of DHS and submitted to another part of DHS for approval. All urban areas have to submit these strategies.

Chairman TOM DAVIS. So you were basically carrying out in your grant application what the Department of Homeland Security had asked you and worked with you to do.

Mr. REISKIN. Absolutely. We followed the guidance to the letter from the Department.

Chairman TOM DAVIS. Did they at any point say, we don't think this is going to cut it? We think you need to do more? As you were working through the grant process, did they say that anything was remiss?

Mr. REISKIN. We did not get feedback from the Office of Grants and Training.

Chairman TOM DAVIS. Did you fully understand how the DHS Homeland Security grants evaluation process was different for fiscal year 2006, that you would have to first submit fully justified plans for evaluation before any funds were allocated to your jurisdiction, unlike previous years with where you were awarded an amount without a detailed explanation?

Mr. REISKIN. No, actually the Department was fairly clear at a high level about the risk, the effectiveness, the fact that we were now competing for these funds, the fact that we would have to justify our need. We didn't have visibility, and I'm not sure they even had it developed at the time what the actual formulas would be, how certain things would be weighted one versus the other, but they were fairly clear that they were using this risk.

Chairman TOM DAVIS. Different than it had been before?

Mr. REISKIN. That it was very different. To their credit, I think they did a fair bit of outreach and documented process to make that clear.

Chairman TOM DAVIS. Did you make adjustments in the application preparation process from what you had in previous years?

Mr. REISKIN. It was almost unrecognizably different.

Chairman TOM DAVIS. Did you undergo any internal evaluation to determine the best strategy for preparing and submitting the application in light of the changes at DHS?

Mr. REISKIN. There was considerable discussion among the State level group and local level group in addition to the emergency preparedness council for the region which has local elected leaders, private sector folks, nonprofit folks. We discussed starting early in December when the guidance was first released.

Chairman TOM DAVIS. Mr. Foresman, let me just ask you, the city's application was not rated that highly. When you look, it was in the bottom quarter, I think. They were coordinating with the local Homeland Security folks. As he said, he ran this up the ladder, and it had kind of been approved. I don't expect DHS to try to grease the wheels for the city. I think it is a competitive process. But in light of that, what was deficient about their application when you compare it to other jurisdictions?

Mr. FORESMAN. Congressman, I don't have the specifics on the National Capital Region, and we will provide that, but let me be——

Chairman TOM DAVIS. Why not? You knew you were coming here today.

Mr. FORESMAN. Mr. Chairman, let me get into——

Chairman TOM DAVIS. You knew you were coming here today to talk about this.

Mr. FORESMAN. Yes, sir.

Chairman TOM DAVIS. I think they're interested in what they didn't do right. They sought guidance from DHS as they went through this process. I think there was every expectation that they wouldn't be cut 40 percent. I'm just trying to ask what went wrong here. Can you give me some help?

Mr. FORESMAN. Mr. Chairman, absolutely. Let me be very clear, they were not deficient in their application, and I very much want to underscore the nature of the peer review or the peer preparation process that they used here in the National Capital Region.

In the context of the effectiveness scoring, the effectiveness measures, their application was reviewed in relation to the others submitted from urban areas around the country, a relative ranking provided to that. And the simple reason for doing this, Mr. Chairman, is to encourage regions to submit applications that are consistent with the national priorities that will be sustainable even beyond the end of Federal grant funding.

And in terms of their ranking, it doesn't mean that they did a good job or bad job on writing the application; it simply means that, in terms of where they chose to make their investments in relation to the other urban areas around the country, there was some level of ranking.

The challenge, Mr. Chairman, I would have had coming in here is I would have had to have to know the specifics of what was in 456 investment justifications submitted from around the country and the specifics for all of your members on your committee. So I will be more than happy to provide you a very detailed written response, sir.

Chairman TOM DAVIS. Are you comfortable with the evaluation criteria that DHS used this year to come to these outcomes, or do you suspect maybe next year, in light of some of the criticisms this has drawn, the way the allocation process works, that maybe it ought to be reviewed?

Mr. FORESMAN. Congressman, Mr. Chairman, the one thing I will tell you is that I never let any program sit still without seeking continuous improvement. That has been the hallmark of the way that I have approached public policy over nearly a quarter century of service at the State and local level.

I will tell you, though, that I feel confident in our risk analysis process. I feel confident in our peer review process, but we have not been effective in articulating and communicating that as we should have to the stakeholder community including the U.S. Congress. But, Mr. Chairman, if we can find something we can do better next year, we will absolutely do it.

Chairman TOM DAVIS. George, I have a high regard for you. You were great in Virginia, and you have a tough job here, and you're defending something here that you didn't do these decisions, were peer review work. But I have to tell you, on the risk analysis side, if you're comfortable on the State side comparing the Nation's Capital to Montana and comfortable putting Montana and North Dakota and South Dakota ahead of the District in terms of the risk analysis as you defined it, I think there is something wrong.

Mr. FORESMAN. Mr. Chairman, let me be very clear on that particularly. I was referring to the risk analysis as we did the threat vulnerability and consequence piece of it. The one thing to understand about D.C.'s ranking in relation to Montana is that is based on a formula set by the Congress that in large part drives our funding allocation process there, and it drives the way that we analyze this.

I'm not comfortable, frankly, comparing the District of Columbia to 50 other States. I think that's unfair to the District, but it is the nature of the program. It's the nature that they're in the program, and so we've got to make the best of it.

Chairman TOM DAVIS. They were ranked below every State, weren't they?

Mr. FORESMAN. Primarily because of the population issue.

Chairman TOM DAVIS. I think they've got more people than Alaska.

Mr. FORESMAN. I don't know that they were at the total bottom.

Chairman TOM DAVIS. I think there were some territories that ranked ahead of it.

We will look forward to that information, but when you are taking a look at your grants and you have a factor of five for the State grants and seven for the urban grants, I think it really skews the grant process because vulnerable areas like the District and New York end up on the short end of the stick.

Go ahead.

Mr. FORESMAN. Mr. Chairman, I would just offer, I think that is indicative of the discussion that we have had here this morning, where it's difficult to keep separate the discussion about the State grant programs and the urban area grants programs, how risk versus a formula plays into each one, and I think it underscores how we need to do a much more effective job in our communication with the State and local partners in future years.

Chairman TOM DAVIS. Let me ask this, under DHS funding requirements, a program and capability review is the first step in preparing a program capability enhancement plan, which is the building block for preparing investment justifications for Federal funding. What approach did the National Capital Region use in its program and capability review?

For example, did the National Capital Region use a risk-based approach based on assessing current capabilities, determines how these capabilities responded to regional risks and threats and the gap between current and needed capabilities? What key information pieces were used in the review? Did the NCR review individual jurisdiction plans? Can you help me with that?

Mr. REISKIN. The first thing we did was we availed ourselves of technical assistance offered by the Department of Homeland Security to kick off the program and capability review process both for the District and for the region. As I mentioned, it's outlined in our testimony as well, we identified six target capabilities in addition to the eight national priority capabilities. We then basically held workshops with the experts in each of those 14 capability areas for the region. Again, these are the fire chiefs, the police chiefs, health directors. We used that workshop process, following the guidance religiously, to work through the identified strengths, the weaknesses, things that were identified through our strategic planning process last year; to the extent that we had the information, threats that we faced, and out of that developed a pretty comprehensive plan in the areas of people, equipment, training, other resources, fairly complete and voluminous, of strengths and weaknesses for each of the 14 capability areas.

It was out of that we developed the investment plans, out of that we developed the investment justifications that we submitted in our application.

Chairman TOM DAVIS. Let me ask you this question. It's not quite a friendly question, but can I get the answer because, according to research by the American Enterprise Institute in March 2006, "Washington, DC, used the region's first wave of Homeland Security aid as seed money for a computerized car-towing system, and the city also used $100,000 in homeland security money to fund the popular Summer Jobs Program." Is that accurate?

Mr. REISKIN. These are not new issues that arose. This is not Department of Homeland Security funds. This is the $168.8 million congressional appropriation we received in fiscal year 2002. There was——

Chairman TOM DAVIS. So that wasn't this kind of grant money or earmarked Homeland Security money?

Mr. REISKIN. This preceded the standup of the Department of Homeland Security. That was out of a congressional appropriation. The Summer Jobs Program, I believe, was on the order of $18,000 out of the $168.8 million allocation, and there was a program, it's called—we actually still have it through our Department of Employment Services, called Team D.C., where we engage youth in disaster preparedness and community outreach surrounding disasters, and it's a program that we actually think is very good.

The computerized towing contract, I can't say for certain if that was funded. I think it was $35,000 out of $168 million. If it was, I would offer that the ability to efficiently move stalled vehicles during a disaster would not have been an unwise expenditure of funds.

Chairman TOM DAVIS. I just wanted to give you an opportunity to put it on the record.

Ms. Norton.

Ms. NORTON. Just a few more questions, Mr. Chairman.

I just want to understand, Mr. Foresman, whether your testimony is that Congress, it is Congress that mandated the increases. For example, that Louisville got a 70 percent increase, that Omaha got a 61 percent increase, you're saying Congress is responsible for that.

Mr. FORESMAN. Congresswoman, what I am offering to you is the congressional direction was to move to a risk-based process and an effectiveness scoring system, and those things contributed to the changes. And the percentages, while dramatic, represent——

Ms. NORTON. Congress only mandated a risk-based system. You had to then go and figure out what that meant. Congress doesn't have anything to do with these increases or with these cuts.

They're based on the formula that the Department came up with, not with any, "congressional mandate."

Of course everybody on this panel is for a risk-based system; the controversy is about whether we mean the same things when we say that.

Mr. Foresman, when I hear you describe the competition, and we value nothing more than the competitive process of the Federal Government, but when I heard you discuss the competition for these funds that you couldn't tell the chairman until you looked at all the other ones, etc., it sounded like nothing—it sounded exactly like the kind of competition that States and localities go through when they're applying for education grants and the rest.

It was kind of scary to hear that it was the kind of same old competition, albeit with subsections, just like other competitions for grants have subsections.

The reason I ask that or raise that is, you are, of course, aware that this region, unlike any region in the country, by mandate of Congress in the Department of Homeland Security legislation, has a coordinator for this region paid by the Department, that is to say by the Federal Government.

I would like to know, in light of the fact that Congress—you want to know what Congress has mandated, when it lays out the money for a Federal coordinator and says that it will be paid out of DHS funds, Congress is mandating something very specific.

I want to know whether or not, in light of this unique role, whether the coordinator was given instructions about how he was to relate to this region so as to make sure that, in his paid capacity, he helped the region to make out the kind of application that could protect all the Federal assets here. What was his role?

Mr. FORESMAN. Congresswoman, there are two parts to the answer here. First, Tom Lockwood, who serves as our Director of the Office of National Capital Region Coordination, the National Capital Region is the only place where we have a single, full-time dedicated office and Federal personnel dedicated to regional activities. We are moving to a regional structure elsewhere in the country, but here in the National Capital Region, Tom works a wealth of issues from working with the communities on strategic planning preparation, operational coordination, coordination among the Federal work force issues.

Ms. NORTON. I want to know, given the importance of this region, that we are paying somebody to make sure that this region, even if the rest of the regions fall flat, that this region does what it is supposed to do because of what is located here. I want to know whether he had a specific role in making sure that these applications met your terms?

Mr. FORESMAN. Congresswoman, I will answer it this way: Mr. Lockwood works with his three counterparts on a regular basis——

Ms. NORTON. Yes or no? Did he have a role? My goodness, Mr. Foresman.

Mr. FORESMAN. He was involved in the strategic planning process, Congresswoman.

Mr. NORWOOD. He did not have a role, it seems.

Did he have a role, Mr. Reiskin? Did he have a role versus the panel? Did he tell you, look, this is the National Capital Region,

I am the coordinator; let me just put you on notice, these are the kind of things you have to do in order to get your applications to pass muster? Could you all tell me, please?

Mr. REISKIN. I would say that Mr. Lockwood in his capacity serves as a peer with us on the senior policy group which makes the final decisions in terms of the submission of the application and the ultimate award of the funds. He was not serving as a kind of liaison to the Office of Grants and Training, which was driving the grant funding process. I think they might argue it would have been a conflict for him to do so.

Mr. NORWOOD. Would it have been a conflict? That is really my point. If you were only protecting all of us, that might have been a conflict. This Department, this Congress and every Federal asset, which happens to be virtually all of them in the United States, are implicated here.

I am not talking about whether or not he was helping you to protect D.C. or even Maryland and Virginia. I am talking now about, and I have to ask you, Mr. Foresman, if something happens here, are you prepared to say, for want of a good application, they didn't get the funds?

We are thinking about the uniqueness of this region, and I really don't hear in your discussion of the competition a sensitivity to that uniqueness as indicated by the Congress with a Federal coordinator paid by Federal funds.

It sounds as though this was a region among others. We looked at it as we look at others. We will look at 4,000 other applications, and then we will see how you came out. If that is what happened, that raises a very serious question.

For example, let me ask you this: We know that asset risk is generally generic. Here I am trying to point to the difference between this region and many others.

However, let's take Union Station. If we are talking about an ordinary train station, everybody has a train station. Philadelphia has a train station. Every place has a train station. This is what we are talking about.

We are talking about the hub that the Federal workers use as a transportation hub for the region. We are talking about the Union Station that is within a couple of blocks of the Capitol. We are talking about Union Station where Members of Congress have events on the premises and can hear the trains rolling underground as they come in.

In calculating the risk of Union Station, do you in fact treat it as other rail stations, or how do you in fact calculate the risk in this example I pose?

Mr. FORESMAN. Congresswoman, let me go back to the first part of your question where you talked about the effectiveness score. I want to be very clear that 66 percent of the funding is based solely on the risk piece of it; 33 percent is based on the effectiveness.

Ms. NORTON. Say that again?

Mr. FORESMAN. Sixty-six percent of the funding allocations are based solely on risk. It has nothing to do with effectiveness. And of the 33 percent that is based on effectiveness, the maximum, the maximum that any one jurisdiction could be affected by that effectiveness score is 7 percent of their total allocation.

Now, to the second part of your question——

Ms. NORTON. That wasn't my question. I asked about Union Station.

Mr. FORESMAN. No, you asked me a question, and the first part, I was trying to address that one for you as well.

Now with regard to Union Station, when we look at a specific asset, we look at it in the context of a particular geographical area, and in this case, the National Capital Region is a unique geographical area. We give special attention to the National Capital Region because it is the seat of government. So in the context of our evaluation, we take into account that it is not your average train station.

Chairman TOM DAVIS. Thank you.

Mr. Van Hollen.

Mr. VAN HOLLEN. Thank you, Mr. Chairman. A lot of the areas have been covered here, but I do want to ask you, with respect to the amount of funds, Mr. Foresman, for the D.C. area under the Urban Security Initiative, as I understood your answer to the last question, even if the D.C. area had gotten the maximum on the effectiveness score, the total increase would have been 7 percent? I just want to make sure I understand.

Mr. FORESMAN. Give or take 7 percent.

Mr. VAN HOLLEN. But you graded their submission, as I understand it, in the bottom 25 percent on effectiveness, is that right?

Mr. FORESMAN. Congressman, I will need to double-check to see if they were in the bottom 25 percent of effectiveness, but I believe that is correct.

Mr. VAN HOLLEN. I guess I really haven't heard yet today why it is they scored in the bottom 25 percent effectiveness in your opinion. We have representatives here who were very involved in the submission. We have heard now that Mr. Lockwood, I guess, was not involved, and I want to ask a couple of questions about that. But we have representatives here who know their stuff; they know their business. They have teams of people working on it. Why was their application graded in the bottom 25 percent? The bottom 25 percent in my kids' school is a failing grade. I would like to know why that was. They would like, I am sure, to know, so if they disagree with you, they can contest it; or if they agree with you, they can improve in the future.

Mr. FORESMAN. Congressman, let me be very specific. This was not a grade. This was a relative measure to all of the other urban area investments that were submitted.

Mr. VAN HOLLEN. Where you graded on a curve.

Mr. FORESMAN. In the context of it, from the standpoint of those urban areas, they submitted proposals which were deemed during the review process by the reviewers to be at a certain level of compliance in terms of sustainability beyond the end of the grant program, consistent with the philosophical approach of the national program, and this goes back to the New York piece of it, Congressman. In the context of it, we had this discussion with Mayor Bloomberg, they had a different philosophical approach than what the national program is for.

We are trying to develop a bunch of independent urban area activities as part of an interdependent national approach, so we have to have some target from a national perspective.

But I want to be very clear in saying, having worked closely with these gentleman over the years, having worked in the National Capital Region and living here today, I am proud of the work that has been done. I feel safe in terms of what they are doing.

This was not a scoring of the capability of the National Capital Region. This was a ranking in terms of how well the investment justifications comported with the national priorities in relation to the 400 or so that we received from other metropolitan areas.

Again, I go back to the scoring piece of it. At the end of the day, there has been much made about this effectiveness score. In the most extreme case, the maximum of adjustment would have been about 7 percent. I don't know in the context of where they fall what that would have done to the dollars, but we are talking in, probably, the single digit millions here in the National Capital Region.

Mr. VAN HOLLEN. But let me just get to that issue. You mentioned sustainability and lack of compatibility with the national plan. If you could just be specific about it.

Mr. FORESMAN. Yes, sir. One of the things that Congress directed that we do is to develop the National Preparedness Goal to make sure that we were providing States and communities and, frankly, the Federal interagency with a clear road map of what levels of preparedness are we trying to get to in urban areas, in States, in regions, in individual communities; what do we need to do as a nation?

So the National Preparedness Goal informs the grant programs in terms of the priorities in those grant programs. So what we are essentially saying to a particular community, or in this case an urban area, how will your proposals increase your ability to get to that desired National Preparedness Goal and also fit within your regional strategy and be sustainable beyond the end of the Federal grant funds?

Congressman, let me be clear. This is a tough position for all of us. But, you know, at the end of the day, the Federal assistance is not designed to be the primary tool for protecting communities. It is designed to supplement local and State dollars that this region has been very good about committing to its public safety activities, whether crime or terrorism.

And, you know, I will tell you the challenge that we face from a public policy debate is we have not had to do give and take at the State and local level in terms of what we are not going to do. If we feel like these security measures that aren't being funded by the Federal Government, if they are that important, are we still going to go ahead and do them?

Mr. VAN HOLLEN. I appreciate the fact that the local jurisdiction is supposed to be the primary mover in this area, and I think all of these gentlemen next to you do as well. But that doesn't mean that it is not important for us to rely on the Federal contribution, especially in the Nation's Capital. So we are all still struggling I think with why we are seeing such a reduction from last year.

Part of it, obviously, is due to the reduction of the overall level. Part of it may be due to risk adjustments upward for other areas.

Part of it, though, is due to the fact that, at least on a curve, we were at the very bottom.

I guess if I could just ask the other witnesses at the table whether you have had an opportunity to look at the criticisms or the evaluation by the Department of Homeland Security and whether you have any responses to them or whether you feel that you are still left in the dark as to why you didn't measure up relative to other jurisdictions that submitted their proposals?

Mr. CROUCH. I will respond to that, Congressman. Thank you.

We have been offered opportunities by the Department of Homeland Security, including Under Secretary Foresman, to meet with them in the future to get that detailed analysis. We are very eager to do that for all the reasons you so well stated just awhile ago.

Mr. VAN HOLLEN. Let me make it clear. You have not yet had an opportunity to review any of that?

Mr. CROUCH. We have not.

Mr. VAN HOLLEN. One last question that goes to Mr. Lockwood's role. I think we would all agree, Mr. Foresman, I am sure you would agree, the goal is to have in place a plan that best protects people, the National Capital Area and other areas of threat around the country; right?

Mr. FORESMAN. Yes, sir.

Mr. VAN HOLLEN. So I am a little concerned about this idea that Lockwood's participation in this effort to come up with the plan would somehow be seen to give this area a disadvantage, because our goal should be for him to provide the input that makes it the very best plan we can have so that the people in this area can access the full amount of resources that are potentially available for them to access.

This idea that somehow when we, the Congress, have created this position specifically because it recognized the kind of unique threat posed to people in the National Capital area, to say that person would somehow be, you know, it wouldn't be fair for that person to participate, seems to me to put sort of bureaucratic gaming concerns over the welfare of the people of the region. Because what we want is a good plan that everyone agrees is a good plan.

So why would you not allow him to help participate so that this region could get the very best plan possible?

Mr. FORESMAN. Congressman, I don't believe I said, it wouldn't be fair, and, if I did, I was incorrect in that statement. It may have been said by one of the other panelists. We expect Tom to be engaged, but he has got a multitude of things that he is engaged in. The greater expectation is for our team in our Grants and Training Office, who works with the community on a regular basis, is in constant contact with them, to provide them the advice and counsel.

But, Congressman, let me also be clear that one of the things, and the chairman pointed this out, that this needs to be about a transparent process that is not driven by politics, and there have been some incorrect allegations to that extent. This has got to be driven by a very clear process that puts everybody in an equitable playing field, but doesn't do it in such a bureaucratic way that we don't get the best solution for the region. So we understand that.

Tom is not excluded from the process, but he doesn't have the intimate grant expertise that the other folks who are assigned, the gentlemen who are assigned here in the National Capital Region, will have.

So it may be more appropriate to say Tom was involved to the degree he should have been involved, but we also had our team members from our Grants and Training shop who are working with the individual States and the National Capital Region as a regional jurisdiction, working with them throughout the process.

Mr. VAN HOLLEN. Thank you, Mr. Chairman.

Chairman TOM DAVIS. Mr. Fossella.

Mr. FOSSELLA. Thank you, Mr. Chairman.

Mr. Foresman, it is my understanding that the funds from the State Homeland Security Grant Program can be used by the States to increase preparedness for building basic security capacities in their States and the UASI program is intended to protect the largest urban areas in the country. Is that correct?

Mr. FORESMAN. That is correct, Congressman.

Mr. FOSSELLA. So the question I have is whether the States and cities, shouldn't they be competing within the State Homeland Security Grant money for "building basic security capacities" instead of UASI?

Mr. FORESMAN. Congressman, I think in the context of the guidance that we received as part of the appropriations process, they did not make the distinction between the two programs. So what we are looking to do is to build capacity in both contexts.

In the case of New York City, we had the conversation with the mayor, you know, is capacity paying for the overtime costs in the case of the New York Police Department or other things? It represents a philosophical discussion that will go into the grant development process in the future.

Mr. FOSSELLA. So do you see the UASI grant program as an opportunity to shift funds, let's say, to maybe what many of us believe—I don't think it is a philosophical difference. I thought that the State Homeland Security Grant Program is different from UASI insofar as the State Homeland Security Grant Program was to allow States to increase preparedness for building basic security capacities, as opposed to UASI, which was originally intended for the large Urban High Threat Areas.

You are saying you can go in there and take the money out of UASI because you have been guided by the appropriations process to do so?

Mr. FORESMAN. No, Congressman. If that is what you perceived I said, that is not what I intended to say. We view these programs in the context of a singular grant program. What we are looking to do is to make sure that expenditures at the urban area are consistent and not conflicting with what a particular State is doing as part of an overall national initiative.

Mr. FOSSELLA. When you say single, so you see both fund programs as one funding stream?

Mr. FORESMAN. We see them in the context of broad Homeland Security funding but specific different program areas.

Mr. FOSSELLA. You don't sit there and say, OK, we are going to take some money from UASI and some money from there. You see it as one pot of money to comply with your goal?

Mr. FORESMAN. Well, based on the guidance that we are provided by Congress, we have the allocation within the particular sub-program areas, if you will.

Mr. FOSSELLA. Sub-program areas?

Mr. FORESMAN. Between the urban area program and the State Homeland Security Grant Program.

Mr. FOSSELLA. I don't think I have heard you. Do you see it as one pot of money or not? Yes or no?

Mr. FORESMAN. Yes, Congressman, we do see it as one large pot of money, with many different utilizations.

Mr. FOSSELLA. We respect to, among the different submissions, particularly to the purity of process, can you reveal where New York fell within that process on the applications?

Mr. FORESMAN. New York State or New York City?

Mr. FOSSELLA. New York City.

Mr. FORESMAN. Congressman, I don't know, but I will provide you a written response.

Mr. FOSSELLA. It was my understanding it was toward the bottom, like toward the lower, like Washington, DC

Mr. FORESMAN. That is generally my recollection based on the discussion that Secretary Chertoff and I had with Mayor Bloomberg. Again, one of the things, it was a very good, a very constructive discussion in the context of, did New York, from a philosophical standpoint, about 27 percent of what they were asking for was personnel costs, which tend not to be allowed in broad terms under the UASI program. So that was part of it.

There were other issues. But, again, it doesn't point to bad programs in New York City. It doesn't point to bad activities in New York City. It just simply says that, in terms of your relative ranking, they scored differently than the others submitted from around the country.

Mr. FOSSELLA. It was my understanding, again, it was toward the bottom. I think you are confirming that. But you say you are going to get that.

Two points, one on process, one on substance. That conversation took place after the grant program was announced; is that right?

Mr. FORESMAN. That is correct.

Mr. FOSSELLA. Do you think it was the responsibility of anybody at Homeland Security to have an obligation to go back to, for example, New York City and say that your application or your submissions are wrong, rather than wait to make that determination? Do you think anybody had that obligation?

Mr. FORESMAN. Congressman, in terms of the process, there is ongoing discussion regular between New York City and their designated liaison in our Grants and Training shop. When the applications came in, we went through those applications to make sure that all of the component pieces were there so that, if someone left out a piece of paper, they weren't going to be excluded from the review process.

But what we did not do, we did not go in and preview any of the applications to say, this is on point, this is not on point. That is what the peer review process was about. With the recognition that, as we have gone through this process, we have tried to make sure that we treat New York and Washington, DC, with the understanding that they are the highest-risk urban areas, that they are unique in many, many ways, but also put it on an equal playing field with the other urban areas from around the country, because, at the end of the day, the challenge is not to be up here saying that we allowed anything other than a very objective review process to have guided our funding decisions.

Mr. FOSSELLA. Back to the substance aspect of it, I am not so sure on the process of whether—I do believe somebody in Homeland Security had an obligation to step forward. If it was a function of who had the best application, you can make the hypothetical that a city with 100 percent threat that submitted a lousy application, but a city with zero threat who submitted the best application would get the funding. So let me push that a side for a second.

Is there any movement to change the philosophical view? A couple weeks ago, there was a conviction in New York city of a man who plotted to blow up the Herald Square Subway station. That was not done with so much building basic security capacity; that was human intelligence for the most part. That was the 1,000-plus New York City police officers to the tune of $200 million on an annual basis, which is larger than the entire Police Department of Denver, CO, for example.

If we determine that human intelligence is the best way to thwart or prevent a terrorist attempt, don't you think that maybe the philosophy that Homeland Security should change in terms of providing adequate funding to meet and thwart those threats, rather than just building basic security capacity?

Mr. FORESMAN. Congressman, two points to that. First, let me address the first point where you said the effectiveness would have—a real bad application versus a real good application.

I just need to be very clear. It would have made very little impact in terms of the total dollars allocated, in New York City or anywhere else.

The second point to the particular philosophical piece. That is the broader public policy discussion that has been had in the U.S. Congress with States and communities over many, many years about, should the Federal Government provide funds to communities for personnel costs, whether it is personnel costs in this case for cops on the beat doing crime reduction or whether it is personnel costs for law enforcement personnel doing human intelligence. That is part of a broader debate.

I will tell you, on the whole piece of intelligence, as a good example, that is the one area where we have provided flexibility to urban areas in the context of intelligence fusion to be able to use these dollars for some personnel costs, because we recognize, prior to September 11th, that was a capacity that very few urban areas, New York City being one of the exceptions, that was a capability that many areas didn't have.

So we do give flexibility in those particular areas, but not for a broad number of people to be out, "collecting intelligence for analysts."

Mr. FOSSELLA. If I hear you correctly, a city like New York, for example, has that flexibility, to a point?

Mr. FORESMAN. To a very limited point in the context of intelligence and from an analytical standpoint, not a collection standpoint. You would think probably about law enforcement personnel on the street being more on the collection side than on the analytical side.

Mr. FOSSELLA. I know there are other questions. I yield back.

Chairman TOM DAVIS. Thank you very much.

Mr. Ruppersberger.

Mr. RUPPERSBERGER. Well, I think one of the things that is occurring here today is that we have people, and you, Mr. Foresman, are attempting to do your job, and you seem to be very competent. But it seems to me that you are trying to defend a major system that, when you look at the end game, it is just from public perception point of view but more from a reality point of view, giving resources to first responders, isn't really working.

To have a formula that really discounts the threats to Washington, DC, to New York, to those areas, to come up with the way it is, something is wrong.

Homeland Security has had a lot of bumps, and I hope we have learned from those bumps, and I hope that we have become a lot more practical and start focusing on realities, because what I heard here today really disturbs me, and I know you have to take your position where you are.

Let me get into some areas. The first thing, what we haven't talked about here today, in all fairness, from what I understand, and correct me if I am wrong, this program was cut by Congress, the State side, by 50 percent and, on the urban side, by 14 percent. Is that correct?

Mr. FORESMAN. That is correct.

Mr. RUPPERSBERGER. OK. So you are sitting here attempting to defend a system that I think is flawed at this point, but you have less money, and that was because of Congress. So we have to look at ourselves and see where we are.

You also have made a comment about what Congress has told us to do. I am not sure what you mean by that, and I don't know if I have time to get into it.

But what my concern is and what I would like to really determine is where your priorities are. When it comes to funding, it is about priorities.

According to the President, and I believe it, too, we are at war with terror. We are at war in another country, but we also have a terrorism war. We need to do a lot of things very quickly, in Homeland Security and FBI, to find cells, to protect our country. But we know right now that we have been attacked, and we have a lot of intelligence of where we think al Qaeda is going to go.

You made a couple comments, I wrote some notes, about equity and how we need to be equitable. The way I see it, when it comes to protecting our citizens from a terrorist attack, it is not about equity. It is getting resources to the areas which we have chosen, which the President has chosen, what the FBI, NSA, CIA have come together—and that is, by the way, what the Director of National Intelligence is about, to find out where the priorities are and where we are going to put the money. I just am concerned with this matrix, that you have grades where they are.

Now, I am going to stop that with you, because I want to get to Mr. Schrader. I am from the State of Maryland. You put together what you thought was a very good proposal. Let's just talk about the State on a risk basis. You have NSA. You have Fort Meade. You have proximity to Washington, DC. In the region, you have BWI Airport. You have the port of Baltimore, and if you want to get nuclear components in and your target is, say, Washington, DC, the port of Baltimore is the closest port. You have Aberdeen, one of the top testing centers for the U.S. military probably in the country. You have all of these areas here. To me, that is an exposure issue, a risk issue.

Now, I think Homeland Security was doing well as far as prioritizing money to New York, money to Washington. Now we have this new matrix which I perceive is a bureaucratic matrix. I would like to hear from you and go through your process, Mr. Schrader; where you think the priorities are, what you did with respect to your presentation and proposal, and then your anxiety on what occurred and how it occurred.

I would like you to get into an area, too, because I am concerned about a lot of this. Even though Mr. Foresman will not admit it, I think a lot of it has to do with who has the best grant writers and who has the ability to put certain people in certain pegs. That might be OK if you are talking about education funding, health funding, but when it comes to fighting terror, I don't think that is OK.

Could you answer that?

Mr. SCHRADER. Congressman Ruppersberger, let me start by saying that the unfortunate part of all of this is, I think DHS had the right intent when they started this process.

Mr. RUPPERSBERGER. I agree with you.

Mr. SCHRADER. The approach and the concept was good. Quite frankly, the other troubling part of this, and I will get into the direct answer, is that if we push them into changing the process again——

Mr. RUPPERSBERGER. By the way, the reason I want you to answer the question is I want Mr. Foresman to hear this.

Mr. SCHRADER. Yes, sir. It would be very troubling at the State and local level to have this process change completely again next year. The nice thing about what we found is, it seemed to be a good framework on the front end that we can then have a multi-year planning process in place, and that is what we did.

We organized the State into regions. We did regional bottoms-up data. We put these governance groups together to review our statewide program. But our policy in Maryland has been one Maryland in coordination with the NCR, because the NCR is not an operational unit.

For example, we have a fusion center, intelligence gathering in Maryland, that we do for a very limited amount of money through our grant process. It is up in Baltimore, but it covers the entire State, and it supports the NCR. So part of the process here is, we look at the entire State, we have a focus on the NCR, but we also have the central Maryland urban area.

As I said earlier, we had estimated that a reasonable amount of money for the State of Maryland would have been about $35 million to $40 million. We ended up getting about $24 million, which, quite frankly, shocked us.

We did have a very good quality application. So what we were assuming was maybe the risk from the two urban areas sucked off the risk from the State. I am being told now that is not the case.

So the bottom line is, we have been working collaboratively with DHS for over 7 or 8 months now to make sure we put the very best application together. We shared a lot of that information with the National Capital Region. There was a tremendous amount of work that went into the application process. And I think it was all for good intention.

Mr. RUPPERSBERGER. Why do you think you were cut?

Mr. SCHRADER. To be honest with you, I am mystified still.

Mr. RUPPERSBERGER. Why do you think you were cut, based on what you have heard here today?

Mr. SCHRADER. I don't know. I don't know.

Chairman TOM DAVIS. Thank you.

Mr. RUPPERSBERGER. Can he finish answering?

Chairman TOM DAVIS. He doesn't know. He said he didn't know.

Mr. SCHRADER. Let me make it clearer. If the State program was cut in half, we had $20 million last year, we would have expected to get $10 million, just if we did nothing, if we threw something up against the wall. We ended up getting $8 million, which didn't make any sense to us. We worked really hard and thought we would get more.

Chairman TOM DAVIS. Let me just ask, did everybody get less than you thought you would get?

Mr. CROUCH. We did.

Chairman TOM DAVIS. I guess the other question, if I could just tag along, is, what regional Homeland Security programs will now have to be underfunded because of the loss of grant money? If each of you could take a second and give us an idea.

Mr. REISKIN. I will start. As I mentioned, we had identified about $250 million worth of need and submitted an application for $188 million. We didn't expect to get an award of $188 million. But all $188 million represents things we think are important in advancing the safety and security of the region. We now have $46.5 million. So about three-quarters of what we had proposed won't get done.

One major thing I will mention, about $42 million was in the area of interoperable communications. Of that, I think $25 million was to continue the basic infrastructure that we are building, a secure, robust, dedicated network for interoperable communications. That $42 million alone is the entire amount.

Chairman TOM DAVIS. Mr. Crouch, what will be underfunded in Virginia, do you think?

Mr. CROUCH. Thank you, Mr. Chairman.

As Mr. Reiskin indicated, we had identified $188 million in grant programs. We have already gone into a process anticipating that we wouldn't get that amount, certainly, to reduce the amount down to $121. That is pretty much where we are in our analysis at this point. So as Deputy Mayor Reiskin pointed out, we are going to have to look at that $121 million and reduce it to the $46 million.

I mentioned some of the meritorious programs we are looking at earlier, the WMATA operations center backup, the fire assistance for the tunnels, programs to deal with special needs. All of the programs that were submitted had merit to get as far as they did in terms of our submission, and again, they had tremendous stakeholder and local first responder participation in their development.

We will now have to begin the process of determining what programs we eliminate, whether we can partially fund some with anticipation of sustaining them in the future and accomplishing those goals but at a slower pace.

So to identify a specific program at this time, Mr. Chairman, I am not able to do that. But, obviously, we think all of the programs, the projects that were presented, had merit, and obviously, we are not going to be able to proceed with all of those.

Mr. SCHRADER. Three things will be difficult. We have been working closely with Prince George's County to help them with their radio system. It will make it more difficult. Our Critical Infrastructure Resilience Program, reaching out to the private sector and getting the private sector organized in Maryland will probably be affected. Then the last thing, we have a major focus on maritime security in the maritime channels leading up to the Capital Region which would be of concern.

Chairman TOM DAVIS. Thank you.

I am sorry, did anybody else want to add anything?

Mr. ROBERTSON. One addition is that the regional system education campaign I think will be quite vulnerable. We cannot educate our citizens on a cyclical, every 2 or 3 years, basis. We have to do that on an ongoing sustained basis.

Chairman TOM DAVIS. Thank you.

Mrs. Maloney.

Mrs. MALONEY. Thank you, Mr. Chairman.

Mr. Foresman, if I could ask you, did Booz Allen Hamilton give the Department of Homeland recommendations on how to rank urban areas based upon information that they gathered from the peer review process?

Mr. FORESMAN. Congresswoman, Booz Allen Hamilton, just so I am clear, they facilitated the peer review groups. Each one of the investments were reviewed, and then they were reviewed as a package. Booz Allen did the facilitation and rolled up the data and gave it to the Federal team members. But the Federal team members were there with the peer review process. They did the administrative backbone work, if you will, the typing and printing and copying and that type of thing.

Mrs. MALONEY. But they did gather together the information they had from this peer review process and gave it to the Department of Homeland Security; correct?

Mr. FORESMAN. Yes, ma'am.

Mrs. MALONEY. And did the Department of Homeland Security follow the recommendations of Booz Allen Hamilton?

Mr. FORESMAN. They were not Booz Allen Hamilton recommendations. They were the peer review recommendations that Booz Allen Hamilton was responsible for simply documenting as an administrative support function.

Congresswoman, I will go back. I do not believe there were any deviations from the peer review process, but let me provide you a clear written response so that I am very accurate on that.

Mrs. MALONEY. So you believe they did follow the peer review process data compiled and presented to the Department of Homeland Security. May this committee have copies of the data and the rankings that were provided to the Department of Homeland Security from Booz Allen Hamilton that compiled it from the peer review process?

Mr. FORESMAN. Congresswoman, we are in the process of providing Chairman King with a significant amount of information. Let me sit down with the internal folks. The Secretary and I are committed to being as transparent as we can. The only two areas that we are trying to find the right accommodation is the classified information, and of course, Congress knows how to deal with classified information. That is easier.

And then on the peer review process, peer review processes are widely used among Federal agencies. We want to make sure, if there were confidentiality agreements with these folks, that we respect those agreements. But we will seek to get you everything you are looking for.

Mrs. MALONEY. You will provide that information to Mr. King and Benny Thompson of the Homeland Security Committee?

Mr. FORESMAN. Congresswoman, we have already set the first delivery over this past Friday, I believe, the first round of material. But we want to get a whole bunch of different material and provide it. I mean, we do. We have every reason to be very transparent with this process.

Mrs. MALONEY. Was it the peer review process that ranked the applications of D.C. and New York City in the bottom 25 percent?

Mr. FORESMAN. The peer review process provided the relative measures. How they were actually ranked against it was, you had 17 peer review teams, and all of this data had to be corresponded because one peer review team didn't do all of the applications. So just from a simple administrative standpoint, someone had to take the findings from the 17 peer review teams and mix it altogether to create the ranking, if you will, as you describe it.

Mrs. MALONEY. So it was the peer review that ranked New York and D.C. in the bottom 25 percent. So my real question then is, how in the world can they make a determination of the risk and need of D.C. and New York City and others if they, as you testified earlier, do not have access to security clearance and to real-time threat analysis?

Mr. FORESMAN. Congresswoman, I did not testify earlier that they did not have access to it. I told you I would provide to the committee a written response just to be clear on it. What I was offering to you earlier——

Mrs. MALONEY. I was told they did not have access to it.

Mr. FORESMAN. Congresswoman, I want to make sure we are dealing with facts and not what people have been told, because that has led to some emotional debate, and I want to be very accurate with you.

Mrs. MALONEY. I would appreciate it if you would let us know how many had security clearances.

Mr. FORESMAN. Absolutely, but it is not relevant to the evaluation of the effectiveness piece. Let me just be very clear about that.

Mrs. MALONEY. Recently we had a hearing on the 9/11 Commission, and both former Governor Kean and Lee Hamilton, a bipartisan commission, testified that the best way to combat terrorism was intelligence, and that this should be our focus. That is why Congress revamped our entire intelligence system now into the National Intelligence Director. They also testified that the biggest threat that our country faces is a so-called dirty bomb, nuclear bomb, set off in New York City.

My question is, did the peer review council read the 9/11 Commission report? Did they take into account the intelligence efforts of localities which our experts say is the place we should be putting our money and our effort in combating terrorism? Zarqawi apparently was captured through intelligence. We have prevented several attacks that people were plotting in New York based on intelligence. We have an anti-terrorism intelligence center that works on this.

I want to know, in your ratings, did you give a higher rating to the possibility that a nuclear bomb would go off in a particular area and kill hundreds of thousands of people, or how was this decision made? Was that part of the rating, since the experts say that is our No. 1 threat to our country at this point?

Mr. FORESMAN. Congresswoman, two parts to that. First, on the intelligence piece, you know, both the co-chairs of the 9/11 Commission are on point with regard to intelligence, and it is because of that, because we are following their recommendations and other commission recommendations, that we have better intelligence, we have better analytical understanding of threats.

Mrs. MALONEY. Was that part of the formula?

Mr. FORESMAN. That is where I am going to. Congresswoman, it was part of the formula, because in order to get to consequences, we have to understand the consequences of a wide range of plausible scenarios in metropolitan areas.

But let me be clear again: New York City is at the top of the risk ranking. It doesn't own exclusive risks. Chicago, LA, and other metropolitan areas do as well, and they also have active threats. We have actually seen active investigations elsewhere. So, you know, there are statements, and then there are facts, and what we are trying to use are factual data to inform our risk analysis.

Chairman TOM DAVIS. The gentlelady's time has expired.

Mr. Clay.

Mr. CLAY. Thank you, Mr. Chairman.

Mr. Foresman, I have major concerns about fiscal responsibility and funding management in regard to these grants. Can you tell us how your agency is addressing these concerns and what standards have been put in place to ensure proper utilization of funds? And, if you could, do you have any examples of waste and abuse in this grant program?

Mr. FORESMAN. Congressman, my most glowing example is the exact reason that we are here today. It is because we simply didn't provide funds to urban areas based on some formula and send the dollars out the door and say, come back to us and report. We asked them to provide various specific investment justifications of how the dollars will produce tangible results.

I will tell you that I am actually quite proud of the men and women in State and local governments across America. While there have been isolated cases of fraud and abuse, I will tell you the vast majority of expenditures that I have personally seen with these Homeland Security funds have measurably improved the safety and security of America. I will tell you, I think the rule is the vast majority of these funds are used very effectively for their intended purposes.

Mr. CLAY. Thank you.

Secretary Chertoff has testified before this committee that DHS analyzes risk using a three-dimensional matrix. One of our colleagues, Ms. Matsui, wrote to Secretary Chertoff asking about how the Department assessed risk for her district in Sacramento.

Let me quote from the reply she got: The Department is continuing to develop a more robust risk model as it gains the capabilities to increase its knowledge of interdependency, cascading effects and refined data sets. These issues are currently being studied to provide both sector-specific and cross-sector modeling through a suite of consequence analysis and decision support tools.

Can you translate this into plain English for me?

Mr. FORESMAN. Congressman, I can.

The tone and tenor of that response is that—and I think the question may have come for two reasons: one, because Sacramento is one of those areas that was deemed, because of the new relative risk ranking, to go off of UASI after this current year. The second piece, and I am not sure, I haven't seen the specific correspondence, we have had issues with levies up there, and we have gotten some letters about, do you consider natural disasters and terrorism risks at the same time?

But if it is strictly in the context of the terrorism piece, it is because our understanding of the risk nationally has changed in other urban areas. From a relative standpoint, it has put Sacramento a little lower on the list.

If we need to get her a clearer response, I will be more than happy to take personal charge of getting that done.

Mr. CLAY. Thank you very much.

Chairman TOM DAVIS. Mr. Clay, unfortunately, we have to get the bus over to the Pentagon, which leaves in 3 minutes.

Mr. CLAY. I understand.

[The prepared statement of Hon. Wm. Lacy Clay follows:]

Statement
Congresswoman Wm. Lacy Clay
Government Reform Committee
Hearing entitled: "Regional Insecurity: DHS Grants to the National Capitol Area."
June 15, 2006

Thank you, Chairman Davis and Ranking Member Waxman, for holding this hearing today. I would like to thank all the witnesses for attending and I look forward to hearing from you today.

The question we face today is how we protect our citizens across the nation in a responsible and cost-effective manner against potential terrorist attacks.

The Department of Homeland Security though it's grant program has been charged with allocating proper funding and resources to States and urban areas for plans to protect against terrorism and other threats to health and public safety.

It is clear that the National Capital Area is at risk, we all remember the tragic events of September 11th, this region like New York was changed forever. However, we must seek to balance what has already happened with a clear vision of what might happen in the future. This should not be an issue of proving which region has more risk at the expense of another. We need to insure for the sake of all our citizens that we have effective plans and resources balanced throughout risk regions so that we may properly protect the Homeland.

I have crucial concerns about the Department's plan and formulas for allocating funds. The Department has an unfortunate track record of mishandling potential risk and response. We have to look no further than to the agency's response to Katrina and its aftermath.

That is why it is crucial that we receive a detailed explanation and transparency in our examination of the FY 2006 Homeland Security grants.

There has been an outcry from several corridors regarding the grant allocation process. The National Capital Area has raised concerns regarding this process, calling into question the criteria used to evaluate the grant proposals. I look forward to hearing from the representatives from the National Capital Area along with the representatives of the Department of Homeland Security on this subject.

In every community in this county we are concerned about protecting our citizens, my district is no different. I have strong concerns about having the funds to support the first responders on the ground in our region, no different than any other region. Again, I think it is vital that we seek to have a balanced approach that does not focus on one region at the expense of another.

I also have serious concerns about potentially wasteful use of allocated funds. I think it is important that we focus on fiscal responsibility so that funds are managed properly and effectively.

In the wake of the devasting attacks of 9/11 it clear that we must get this right, the well-fare of our citizens depends on it. We must examine if we have enough funding for the true risk areas of our nation, however we must also have standards in place so that we are not wasting or ineffectively utilizing funds and plans.

Thank you Mr. Chairman. I look forward to hearing from the witnesses about how we can be effective, balanced and fiscally responsible in our plans to protect the National Capitol Area and the rest of the country. We must get this right.

I yield back.

Chairman TOM DAVIS. I appreciate it. I hate to cut you short. A lot of this is Mr. Moran's district. Would you yield to him for 1 minute? If you have any questions for the record, you can submit them.

I ask unanimous consent that Resolution 16677, passed unanimously by the Council of the District of Columbia, resolving that Homeland Security funding be targeted to the highest threat jurisdictions be put into the record.

Without objection, so ordered.

Mr. Moran.

[The information referred to follows:]

COUNCIL OF THE DISTRICT OF COLUMBIA
WASHINGTON, D.C. 20004

KATHY PATTERSON
COUNCILMEMBER, WARD 3

CHAIRPERSON
COMMITTEE ON EDUCATION,
LIBRARIES, AND RECREATION

OFFICE: (202) 724-8062
FAX: (202) 724-8118

June 14, 2006

The Honorable Tom Davis
U.S. House of Representatives
2348 Rayburn House Office Building
Washington, D.C. 20515-4611

By mail and facsimile

Dear Congressman Davis:

As you prepare for the Committee on Government Reform's June 15th hearing on Department of Homeland Security (DHS) grants to the national capital area, I wanted to share with you a resolution on this subject that was unanimously approved by the Council of the District of Columbia on June 6th.

The resolution, which I co-introduced with Councilmember Carol Schwartz, calls on the U.S. Congress and DHS to take immediate action to revise the fiscal year 2006 homeland security grant allocations and to reallocate FY 2006 Urban Area Security Initiative (UASI) funding through a formula that truly reflects risk and need. The resolution also calls for changes in the formula for allocating State Homeland Security Grant Program funds to increase funding for the District of Columbia and other high-threat jurisdictions.

As The National Commission on Terrorist Attacks Upon the United States (the 9-11 Commission) concluded in its Final Report, "Homeland security assistance should be based strictly on an assessment of risks and vulnerabilities. Now, in 2004, Washington, D.C., and New York are certainly at the top of any such list ... federal homeland security assistance should not remain a program for general revenue sharing."

I would also like you to know that the Council is considering an omnibus piece of homeland security legislation (Bill 16-242, the "Homeland Security, Risk Reduction, and Preparedness Act of 2005," a copy of which is enclosed) that would mandate a threat and vulnerability assessment for the District and create an expert Homeland Security Commission to advise the District on its homeland security policies and offer an independent assessment. The District remains determined to take all of the necessary steps within its own power to increase the safety of our residents.

Thank you for your oversight of these critical issues for the District of Columbia and the entire national capital region.

Sincerely,

Kathy Patterson

Kathy Patterson

COUNCIL OF THE DISTRICT OF COLUMBIA
OFFICE OF THE GENERAL COUNSEL
WASHINGTON, D.C. 20004

I. TRANSMITTAL OF ENROLLED RESOLUTION

Short Title: Sense of the Council That Federal Homeland Security Funding Must Be Targeted on the Highest-Threat Jurisdictions Emergency Resolution of 2006

Resolution #16-677

PR #16-815

Date Transmitted to LSD:

cc: Councilmember Schwartz (with attachment)

Please notify me of any error/comment within 24 hours.

II. CLEARANCE OF ENROLLED RESOLUTION

Legislative Counsel — Date: 6-8-06

Office of the Secretary — Date:

III. CLEARANCE OF CORRESPONDENCE BY LSD

Letter(s) to the Mayor — Date:

Letter(s) to Congress — Date:

Rev. 1/06

ENROLLED ORIGINAL

A RESOLUTION

16-677

IN THE COUNCIL OF THE DISTRICT OF THE COLUMBIA

June 6, 2006

To declare, on an emergency basis, the sense of the Council that fiscal year 2006 homeland security grant program allocations recently announced by the United States Department of Homeland Security fail to target the highest-threat jurisdictions, including the District of Columbia and the Washington, D.C. metropolitan area; that the announced grant allocations will impede efforts to promote regional coordination while undermining the safety and security of residents and wasting taxpayer dollars; that the United States Department of Homeland Security must take immediate action to revise the fiscal year 2006 homeland security grant allocations through funding formulas that truly reflect risk and need; that the $31 million reduction in Urban Area Security Initiative funding for the Washington, D.C. region must be reversed; and that the State Homeland Security Grant Program must provide more funding for the District of Columbia and other high-threat jurisdictions while reducing funding for low-risk jurisdictions.

RESOLVED, BY THE COUNCIL OF THE DISTRICT OF COLUMBIA, That this resolution may be cited as the "Sense of the Council That Federal Homeland Security Funding Must Be Targeted on the Highest-Threat Jurisdictions Emergency Resolution of 2006".

Sec. 2. The Council finds that:

(1) The National Commission on Terrorist Attacks Upon the United States ("9-11 Commission") concluded in its Final Report that, "Homeland security assistance should be based strictly on an assessment of risks and vulnerabilities. Now, in 2004, Washington, D.C., and New York City are certainly at the top of any such list ... federal homeland security assistance should not remain a program for general revenue sharing."

(2) On December 5, 2005, 9-11 Commission Chairman Thomas Kean and Vice Chairman Lee Hamilton stated that, "It is scandalous that we still allocate scarce homeland security dollars on the basis of pork-barrel spending, not risk."

(3) For fiscal year 2006, the United States Department of Homeland Security had announced plans to allocate funding for the Urban Area Security Initiative ("UASI"), a grant program designed to address the unique homeland security needs of large urban areas, based on a determination of risk and need.

(4) On May 31, 2006, the United States Department of Homeland Security announced the fiscal year 2006 Homeland Security Grant Program allocations, which included a $31 million reduction in UASI funding for the Washington, D.C. metropolitan area, from $77 million to $46 million. The fiscal year 2006 grant allocations also reflected a reduction in UASI funding for high-threat cities such as New York.

(5) The planned reduction in UASI funding for the Washington, D.C. metropolitan area will impede efforts to promote interoperable communications, equipment purchases, training and exercises, and public health preparedness throughout the region, where the challenges of homeland security and emergency preparedness are compounded by the need to coordinate among the District of Columbia, Maryland, and Virginia, as well as 7 counties and 11 cities.

(6) The fiscal year 2006 grant allocations announced by the United States Department of Homeland Security also include $4,270,000 in State Homeland Security Grant Program funds for the District of Columbia, less than the amount awarded to any state. It is unacceptable for the United States government to provide more funding for low-risk jurisdictions than for the District of Columbia, which was targeted in the 9-11 attacks and is one of the highest-threat cities as our nation's capital and the symbol of our democracy.

(7) The fiscal year 2006 grant allocations announced by the United States Department of Homeland Security are diametrically opposite to the risk- and threat-based approach advocated by the 9-11 Commission and many other expert groups and individuals, and would serve to undermine public safety and security while wasting taxpayer dollars.

Sec. 3. It is the sense of the Council that:

(1) The United States Congress and the United States Department of Homeland Security must take immediate action to revise the fiscal year 2006 homeland security grant allocations and to reallocate fiscal year 2006 UASI grant funding through a formula that truly reflects risk and need.

(2) The $31 million reduction in UASI funding for the Washington, D.C. region must be reversed and fiscal year 2006 UASI funding for the D.C. area should exceed the fiscal year 2005 funding level of $77 million.

(3) The formula for allocating State Homeland Security Grant Program funds must be revised to increase funding for the District of Columbia and other high-threat jurisdictions while reducing funding for low-risk jurisdictions.

Sec. 4. The Secretary to the Council shall transmit a copy of this resolution to the President of the United States, the Majority Leader of the United States Senate, the Minority Leader of the United States Senate, the Chairman and ranking minority member of the United States Senate Committee on Homeland Security and Governmental Affairs, the Speaker of the United States House of Representatives, the Minority Leader of the United States House of

ENROLLED ORIGINAL

Representatives, and the Chairman and ranking minority member of the United States House of Representatives Committee on Homeland Security.

Sec. 5. This resolution shall take effect immediately upon the first date of publication in the District of Columbia Register.

Mr. MORAN. I thank my good friend from Missouri and the chairman.

We have had the hearing now, and you have explained and justified and excused your decisions. But what do we do now? Where do we go? Might you consider requesting a supplemental appropriation to address these needs? Are there other programs we could apply to? We need to know more than just an explanation. We need to know, what is the next step that we should pursue? You heard the needs, and I would like to get some guidance.

The reason our first responders are in decent shape in northern Virginia, or at least were for the first couple years, Mr. Davis, Mr. Wolf and I earmarked money. Maybe that is what we need to do in the future, is just earmark the money to make sure it goes to our Nation's priorities.

Can you give us a quick response? What you would do if you were us?

Mr. FORESMAN. Congressman, three quick responses for you. First, we had $7.5 billion worth of requests for a both pot of money that was about $1.7 billion this year. That gives us some measurable level of understanding of the scope of the need out there.

Second, we will hopefully, assuming the budget process goes forward, we will have the 2007 grant programs. I am committed to moving those out the door much quicker and much more effectively than we have done in the past. So I hope to have those out the door, assuming the budget process goes forward quickly, early in the fall.

And there are other grant programs. We will continue to work with the folks here in the NCR to identify those other grant programs that may be able to help them address some of these issues.

Chairman TOM DAVIS. That would be very helpful, if we can look at some of these other programs, given their reliance and the fact this is the first year we changed it.

Mr. Moran, I appreciate the question.

I want to thank you all for being up here.

Mr. Foresman, thank you particularly for being up here and being the chief flak catcher today.

The committee stands adjourned.

[Whereupon, at 12:55 p.m., the committee was adjourned.]

[The prepared statements of Hon. Edolphus Towns, Hon. Elijah E. Cummings, and Hon. C.A. Dutch Ruppersberger, and additional information submitted for the hearing record follow:]

Statement of Rep. Ed Towns (NY-10)
before the Committee on Government Reform

"Regional Insecurity: DHS Grants to the National Capital Area."

Thursday, June 15, 2006
10:00am

Thank you, Mr. Chairman. I am very pleased that the Committee is holding this timely hearing, and I ask the committee's indulgence and that of our guests with regard to the issue as it concerns ***my district*** -- one that must be addressed immediately.

As most of you know, on May 31st the Department of Homeland Security announced funding for its major grant programs for the coming year. Sadly, under its High Threat Urban Areas program, the allotment to my home state of New York was cut significantly (and, I might add, *disproportionately*). While major state and local grant programs were cut from $2.34 billion in fiscal year 2005 to $1.67 billion in 2006 (a 29 percent cut), New York State's allocation was cut by ***36 percent***, from $290.2 million last year to just $183.6 million for FY2006. If funding is to be based on a risk-assessment theory, THIS MAKES NO SENSE.

According to the New York Times, the Bloomberg administration wanted to spend $111 million in the coming year to install a sprawling security camera system in the Wall Street area. To be modeled on London's so-called "Ring of Steel", the plan involved placing video cameras at half of the 535 intersections south of Canal Street and installing surveillance equipment capable of zeroing in on details as small as a license plate. In sum, the city of New York was poised to make the changes needed to further protect our citizens from future attacks. I again mention the "risk-based" theory for funding and my utter confusion regarding the DHS cuts. In my mind, folks, New York is certainly "risk-based".

Regarding the cuts, Police Commissioner Ray Kelly said, "The immediate impact — the *biggest* impact — will be a cessation of the Lower Manhattan Security Initiative. We're going to have to regroup and take a very hard look at that plan." He went on to note that "This program right now has been put on hold until we can take a look at the funding stream."

Before DHS cut our funding, we needed much more money to make the necessary security improvements. Now that we have been short-changed, we are in far greater danger. We need dramatic increases in funding from the Port Security Grant Program, the Rail and Transit Security

Grant Program, and the Buffer Zone Protection Plan grant program. Without it, we are UNPREPARED. Mr. Foresman, please work with those of us in the New York delegation to make us whole. While an ideal DHS package would give us ***more*** money for security improvements, the absolute worst-case scenario is to give us ***less***. I know you stand by me in dreading the thought of another disaster in New York, and I respectfully ask you to do all you can to ensure the safety of all New Yorkers. I look forward to further discussions about this with you and your agency.

Thank you, Mr. Chairman.

U.S. House of Representatives
109th Congress

Opening Statement

Representative Elijah E. Cummings, D-Maryland

Full Committee Hearing: "Regional Insecurity: DHS Grants to the National Capital Area"
Committee on Government Reform

June 8, 2006

Mr. Chairman,

Thank you for holding this vital hearing to examine the Department of Homeland Security's (DHS) funding of the National Capital Area.

I had the unprecedented experience, as you did, Mr. Chairman, of being a Member of this House of Representatives on September 11, 2001.

That day of horror and suffering awakened the American people to the ruthlessness of terrorism, demonstrating how truly vulnerable we all were. Those of us in government made a commitment to do all that we could to prevent future attacks—by securing our homeland.

Now, more than five years after that horrific day, we are falling short on our promise.

The president established DHS, and we agreed to fund it, to address the threat of a terrorist attack on our soil. But since its inception, the department has been plagued by controversy and criticism.

Those of us in Congress hold some responsibility for the problems that exist. Rather than allocating funds in an effective and efficient way, we have politicized the process.

Every city and small town, it seems, wants its piece of the homeland security pie. This would be fine if resources were unlimited. But we know, Mr. Chairman, that they are not. To the contrary, Congress this year allocated fewer, not more homeland security dollars.

In fact, the department took a 25 percent cut, making the process by which it divides funds all the more vital.

In the past, DHS was criticized for distributing funds too liberally, paying for projects and efforts that, while good-intentioned, were not necessarily the best use of resources. So this year the department took a different approach.

It came up with a new formula that, we were told, would better allocate resources to programs based on risk, need and effectiveness. Unfortunately, Mr. Chairman, the specific details of that formula have not been made available to us, so we have no way of evaluating it.

But we can make an assessment of the formula based on the information that we do have.

The department recently announced its funding streams. The decisions that were made, I believe, are troubling to say the least.

We are here today to examine why, according to DHS, the National Capital Region, which includes the District of Columbia, Virginia and my home state of Maryland, deserved a 40 percent cut, while small cities such as Louisville, Kentucky and Omaha, Nebraska saw big increases.

I do not mean to downplay the importance of security measures across the homeland; however, Mr. Chairman, I think you and I can agree that certain regions—specifically those that have been attacked in the past—are more vulnerable than others.

We have been told that programs in the National Capital Region were cut because they are inefficient.

I am the first person to advocate for cleaning up inefficiency in government, but how can we fix these programs without details on why DHS disapproved of them?

I hope that today's hearing will help clear up some of these questions.

I look forward to the testimonies of today's witnesses and yield back the balance of my time.

WITNESSES

The following witnesses are expected to testify:

The Honorable George W. Foresman, Undersecretary of Homeland Security for Preparedness, U.S. Department of Homeland Security

The Honorable Anthony Williams, Mayor, District of Columbia

Mr. Edward D. Reiskin, Deputy Mayor for Public Safety and Justice, District of Columbia

Mr. Robert Crouch, Assistant to the Governor for the Commonwealth Preparedness, Commonwealth of Virginia

Mr. Dennis Schrader, Director, Office of Homeland Security, State of Maryland

Mr. David Robertson, Executive Director, Metropolitan Washington Council of Governments

- Mr. Foresman, DHS has been criticized for politicizing the process when it comes to handing out money. Small towns and cities are consistently receiving funds that might be better used in areas that are more likely to be attacked. My understanding is that the new formula DHS created was intended to make the process less, not more politicized. How does it work?

- Mr. Foresman, I think reasonable people can agree that the National Capital Region and New York City should be priorities for counter-terrorism efforts. How does DHS justify cutting funding for these regions, while increasing funding in places like Louisville, Kentucky and Omaha, Nebraska?

- Mr. Williams, according to recent media reports, the DHS has not been entirely clear about what was wrong with the D.C. programs it chose to cut. Is it clear to you, from your conversations with the department, what you were doing wrong and are you prepared to fix those problems?

- Mr. Williams, Mr. Crouch and Mr. Schrader, DHS cuts to the National Capital Region will clearly have an impact. The District received $4.3 million, down from $9.2 million last year; Virginia received $16.8 million, down from $34 million; and Maryland received $24.3 million, down from $39 million. How will these cuts, in real terms, affect your ability to secure your states?

Congressman C.A. Dutch Ruppersberger
Government Reform Hearing
Regional Insecurity:
DHS Grants to the National Capital Area
Opening Remarks
06.15.06

Thank you Mr. Chairman.

I represent Maryland's 2nd Congressional District which includes Baltimore and part of the National Capital Area. The Port of Baltimore, NSA, BWI airport, Ft. Meade, Aberdeen Proving ground and Sector Baltimore are all in my district.

In 2006 the Department of Homeland Security cut National Capital Region funding 40% from the previous year. They also cut 18 million dollars in homeland security grants for state of Maryland.

It is of great concern to me, the Congressman from Maryland, that DHS would cut counter-terrorism money for New York City and the Washington region which together have been 100% of Al Queda's terrorist attacks on America's soil. Each jurisdiction has been cut by at least 40%.

It is also of great concern to me, that DHS has shown by their actions that Nation's Capitol is a "low-risk" city.

When you compare the State of Maryland with nearly 5 ½ million people, the per-capita spending for DHS dollars is $4.5. That means each person in Maryland gets $4.5 dollars in DHS money. Wyoming received nearly $16.00 per person in DHS funding. Wyoming gets more than 15 times many DHS dollars than they have people – $7.6 million dollars and only 493,000 (four hundred, ninety- three THOUSAND) people!

I like Wyoming, the Grand Tetons, it is one of our most magnificent rural areas in this country. I have been there I've observed grizzly bears and other wild animals. Based on this new funding system it seems that DHS is protecting the grizzly bears in Wyoming over the people who live and work around Ft. Meade and NSA. I choose **people** over bears.

I'm willing to listen, but I find it hard to understand why these high priority targets were given less money than last year.

If DHS has a better understanding of where the risk is, why was the funding for such densely populated urban areas like Washington D.C., New York and Baltimore cut so much?

The new DHS system which supposedly "balances" risk and effectiveness places more emphasis on the quality of the writing of the grant proposal than the actual risk a community faces.

I agree that giving money to states without having a plan is putting the cart before the horse, but giving away Homeland Security dollars Federal on the basis of the quality of how a proposal is written is trying to drive the cart WITHOUT a HORSE.

If the state plans are being analyzed, where is the Congressional oversight? I question whether a group of so-called "peer review experts" should be telling first responders what projects have value and which don't, and determining how much money they should receive. Have these peer review experts ever managed a federal grant or a federal emergency?

This is process is not acceptable and I believe that the new DHS system puts our most vulnerable cities at risk. I think that DHS has put the needs of a few over the needs of the most vulnerable.

I will work with my colleagues to fix the DHS grant formula so it accurately protects America's most vulnerable assets.

Thank you Mr. Chairman

ONE HUNDRED NINTH CONGRESS

Congress of the United States

House of Representatives

COMMITTEE ON GOVERNMENT REFORM

2157 RAYBURN HOUSE OFFICE BUILDING

WASHINGTON, DC 20515–6143

MAJORITY (202) 225–5074
FACSIMILE (202) 225–3974
MINORITY (202) 225–5051
TTY (202) 225–6852

http://reform.house.gov

June 21, 2006

Mr. David Robertson
Executive Director
Metropolitan Washington Council of Governments
Suite 300
777 North Capitol Street, NE
Washington, DC 20002

Dear Mr. Robertson:

Thank you for your testimony before the Committee on Government Reform on June 15, 2006. The Committee greatly appreciated your participation in the discussion about the Urban Area Security Initiative grant process and looks forward to a continued dialogue with the representatives of the National Capital Region about steps being taken to make the capital area safe. This letter sets forth additional questions for the record for the June 15 hearing. Please answer the following questions:

1. The NCR's effectiveness rating was in the bottom 25 percent of all Urban Areas. Reportedly, the proposals were viewed as less innovative and less likely to produce sustained, high-impact results than those submitted by other areas.

   Prior to receiving feedback from DHS on the NCR application (which is scheduled for this month), and based upon further review of the application, what do you see as possible weaknesses associated with the investment justifications?

   What specific changes would you make in any re-submittals of investment justifications?

2. The NCR jurisdictions are currently drafting a NCR strategic plan. Thus, a strategic plan was not available to guide the NCR application for funding.

   To what extent did the lack of a strategic plan impair NCR's preparation of its funding application? Please explain.

How will the completed strategic plan provide the needed justification for future federal funding applications?

Additionally, how is this plan different from the Enhancement Plan? The Enhancement Plan lists vague needs such as, "Additional equipment is needed to protect citizens from attack," and "People need to train regularly." How is the strategic plan more specific, and when will it be released? Likewise, did the grant application to DHS clarify generalities such as these? Please provide examples.

3. In your testimony you state that the NCR has "developed emergency transportation plans."

   Do these plans include a plan to evacuate METRO passengers in the event of an emergency (something the NCR Enhancement Plan indicated has not yet been addressed)? If not, why? If so, please include a brief summary of the plans in place.

4. Your testimony indicates the Senior Policy Group and the Chief Administrative Officers will now prioritize needs and develop a list of projects that can be funded at the current award level. When do you anticipate this list being finished?

   Are there contingency plans for the NCR jurisdictions, acting as a body or independently, to address priorities that will not receive federal funding from UASI? If not, why? If so, please include a brief summary of these priorities.

Please respond to this letter no later than 5:00 p.m. on July 7, 2006. Your response and any supporting documents should be delivered to Room 2157 Rayburn House Office Building. If you have any questions about this letter, please contact Wimberly Fair at 202-225-5074.

Thank you again for your time and assistance with these requests.

Sincerely,

Tom Davis

Tom Davis
Chairman
Committee on Government Reform

METROPOLITAN WASHINGTON COUNCIL OF GOVERNMENTS

Local governments working together for a better metropolitan region

District of Columbia
Bowie
College Park
Frederick County
Gaithersburg
Greenbelt
Montgomery County
Prince George's County
Rockville
Takoma Park
Alexandria
Arlington County
Fairfax
Fairfax County
Falls Church
Loudoun County
Manassas
Manassas Park
Prince William County

June 14, 2006

The Honorable Tom Davis, Chairman
Committee on Government Reform
U.S. House of Representatives
2157 Rayburn House Office Building
Washington, DC 20515-6143

Dear Chairman Davis:

On behalf of the Metropolitan Washington Council of Governments (COG) Board of Directors, I would like to thank you for addressing an issue of crucial significance to the National Capital Region.

COG and the National Capital Region Emergency Preparedness Council (EPC) work diligently with the area's jurisdictions to keep area residents and visitors secure, and we remain committed to ensuring that the capabilities we have built over the past four years will continue to keep them safe. However, without the proper level of funding, our ability to further enhance the region's preparedness will be gravely affected.

The Urban Area Security Initiative (UASI) grant program provides local governments with much-needed financial assistance to address the unique planning, equipment, training, and exercise needs of this high-risk urban area. Put simply, we can not afford to slow our efforts in enhancing the region's capacity to prevent, respond to, and recover from threats or acts of terrorism. It is vital that we remain properly equipped to provide much-needed resources and support, from protective gear for our first responders to strategic plans for the region's leaders to use during the event of an emergency or terrorist attack.

The current level of funding is disappointing and threatens our capacity to move forward at a reasonable pace. COG and the EPC wish to assure members of Congress of the following:

- Local governments in the National Capital Region will respond quickly and effectively to emergencies in this area, using our own resources if necessary and in close coordination with state and federal officials.

- This region's leaders have invested homeland security funds wisely. We have purchased back-up sets of personal protective gear for first responders, we have adopted systems that allow those first responders to communicate with each other and regional leaders, we are working to secure and protect our water supply and we have strengthened emergency transportation plans.

777 North Capitol Street, N.E. Suite 300 Washington, D.C. 20002-4290
Telephone (202) 962-3200 Fax (202) 962-3201 TDD (202) 962-3213 Internet http://www.mwcog.org

- Nonetheless, there are tremendous unmet needs in the region. The U.S. Department of Homeland Security (DHS) identified 37 capabilities that need to be addressed nationally, and DHS asked that we focus on eight of those priorities. Through our own rigorous planning process, six additional target areas were added to make a combined list of 14.

- The impact of the cut in our UASI funds will be serious. Simply stated, we cannot achieve the security improvements this Congress values with the proposed funding level.

- Some have mentioned that the National Capital Region's application was somehow insufficient. Although we believe we followed the instructions we received, let us assure you that we will study all the feedback from DHS and that we are determined to strengthen the application. We are just as committed as we were after 9-11, when we were the first region in the country to respond with a detailed planning process.

We generally support the risk-based approach taken by DHS, but the outcomes are surprising, especially when two-thirds of the determination for the UASI grant is based on the assessed level of risk. Since the events of 9-11, our greatest priority has been an intense commitment to the safety and security of the National Capital Region. That priority and commitment remain strong.

Sincerely,

Jay Fisette, Chair
COG Board of Directors

Gerry Connolly, Chair
National Capital Region Emergency
Preparedness Council

UASI Spending Pressure in the NCR

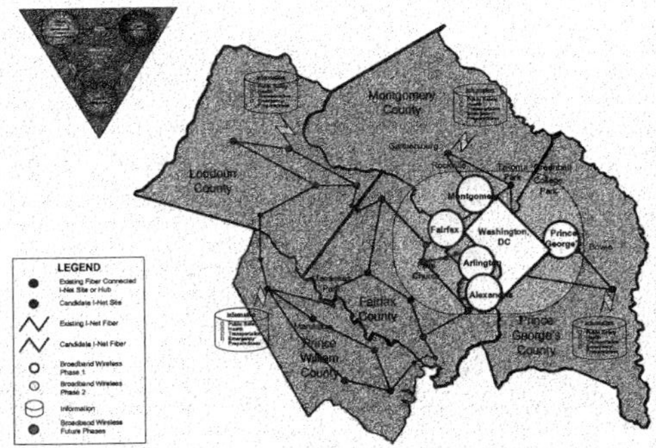

NCR Interoperability Phase 2	$20M
GIS (Regional Incident Maps) Phase 1	$1M
Mass Notification Phase 2	$900K
Prince George's Co. Public Safety Radio System Phase 2	$3M
WMATA Public Safety Radio Communications Upgrade Phase 2	$6.1M

$31M

Interoperable Voice & Data Communications

OR

Critical Infrastructure	$ 18M
Citizen Preparedness / Protection	$12.3M
Critical Resource Logistics	$11.9M
Explosive Devices	$8M
Intel/Info Sharing	$4.1M
Law Enforcement	$7.7M
Mass Care	$5M
Medical Surge	$5.4M
Planning	$10.7M
WMD/HazMat	$6.6M

$89.7M

Other Critical Investment Areas

NCR Regional Data Communications

Historical State

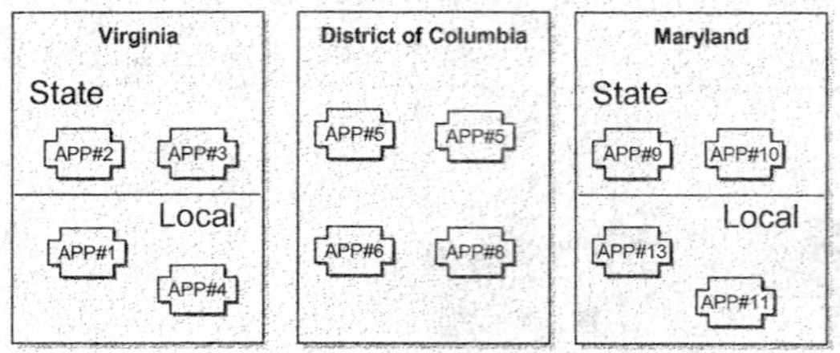

- Independent jurisdiction INETs
- Non-interoperable commercial mobile data
- Limited regional, intra/interstate data interoperability

Future State

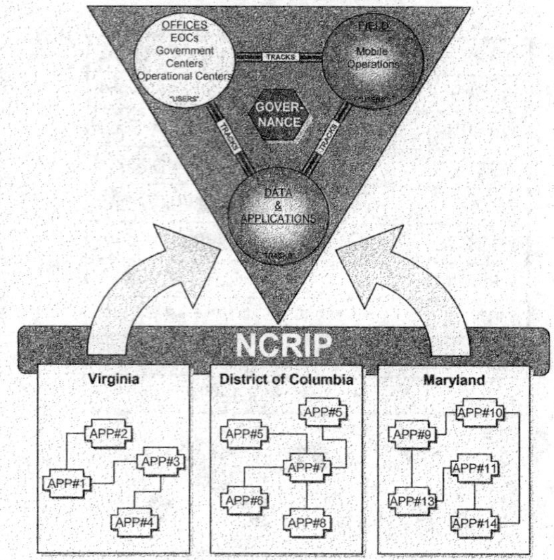

- Integrated regional INETs
- Integrated regional mobile network
- Interoperable regional data

Provides a platform that scales to provide seamlessly interoperable video, data, and later voice communications.

NCR Regional Data Communications

Future State

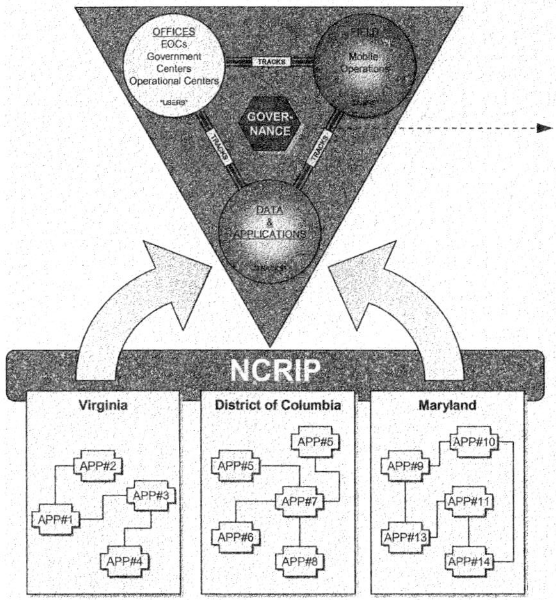

- Executive governance and management organization
- Standard policies and procedures
- Standard technology protocols
- IT security architecture
- Data sharing protocols
- Standard operational protocols
- Operational organization and locations
- Maintenance and sustainability structures
- Strategic/tactical financial models

- **Integrated regional INETs**
- **Integrated regional mobile network**
- **Interoperable regional data**

An infrastructure investment resulting in substantial regional operational cost savings

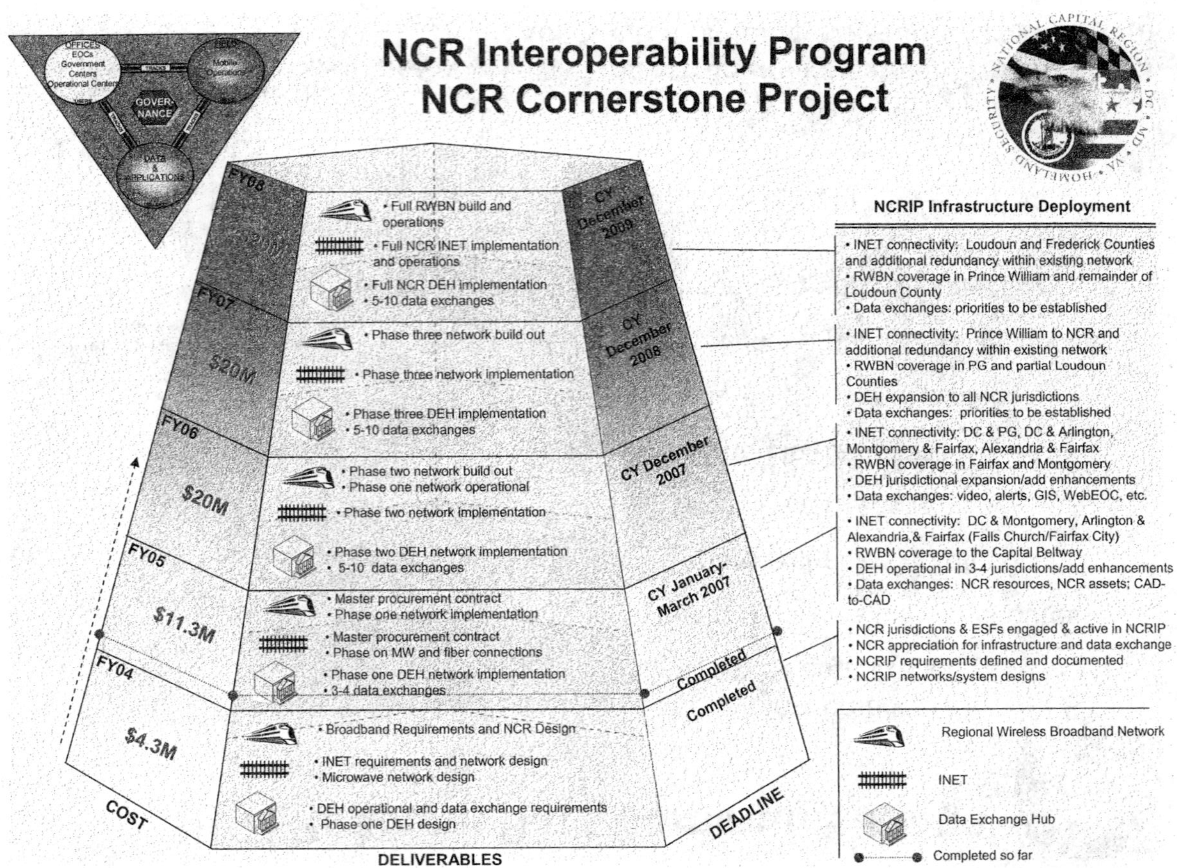
NCR Interoperability Program
NCR Cornerstone Project
OFFICES
EOCs
Government Centers
Operational Centers
GOVER-NANCE
DATA & APPLICATIONS
FY08
FY07
$20M
FY06
$20M
FY05
$11.3M
FY04
$4.3M
COST
• Full RWBN build and operations
• Full NCR INET implementation and operations
• Full NCR DEH implementation
• 5-10 data exchanges
• Phase three network build out
• Phase three network implementation
• Phase three DEH implementation
• 5-10 data exchanges
• Phase two network build out
• Phase one network operational
• Phase two network implementation
• Phase two DEH network implementation
• 5-10 data exchanges
• Master procurement contract
• Phase one network implementation
• Master procurement contract
• Phase on MW and fiber connections
• Phase one DEH network implementation
• 3-4 data exchanges
• Broadband Requirements and NCR Design
• INET requirements and network design
• Microwave network design
• DEH operational and data exchange requirements
• Phase one DEH design
DELIVERABLES
CY December 2009
CY December 2008
CY December 2007
CY January-March 2007
Completed
Completed
DEADLINE
NCRIP Infrastructure Deployment
• INET connectivity: Loudoun and Frederick Counties and additional redundancy within existing network
• RWBN coverage in Prince William and remainder of Loudoun County
• Data exchanges: priorities to be established
• INET connectivity: Prince William to NCR and additional redundancy within existing network
• RWBN coverage in PG and partial Loudoun Counties
• DEH expansion to all NCR jurisdictions
• Data exchanges: priorities to be established
• INET connectivity: DC & PG, DC & Arlington, Montgomery & Fairfax, Alexandria & Fairfax
• RWBN coverage in Fairfax and Montgomery
• DEH jurisdictional expansion/add enhancements
• Data exchanges: video, alerts, GIS, WebEOC, etc.
• INET connectivity: DC & Montgomery, Arlington & Alexandria,& Fairfax (Falls Church/Fairfax City)
• RWBN coverage to the Capital Beltway
• DEH operational in 3-4 jurisdictions/add enhancements
• Data exchanges: NCR resources, NCR assets; CAD-to-CAD
• NCR jurisdictions & ESFs engaged & active in NCRIP
• NCR appreciation for infrastructure and data exchange
• NCRIP requirements defined and documented
• NCRIP networks/system designs
Regional Wireless Broadband Network
INET
Data Exchange Hub
Completed so far

Regional Wireless Broadband Networks (RWBN) Overview

- The National Capital Region (NCR) plans to launch a regional seamlessly interoperable wireless broadband network of networks.

- The benefits gained from the RWBN includes real-time video from cruisers and helicopters to EOCs and command centers and other units, aerial photography and other GIS applications for high-resolution images, graphics rich web-based information.

- These systems will be built to public safety standards.

- The NCR needs a waiver from the current rules to be able to fully operate the network in the Public Safety 700 MHz wideband spectrum.

Regional Wireless Broadband Network Deployment FY05 to FY08

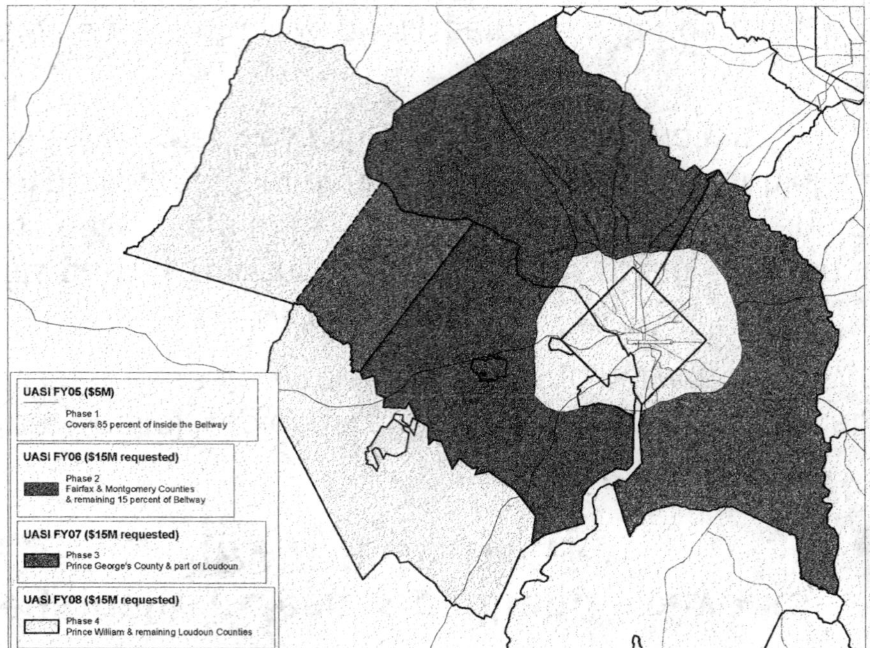

RWBN User Specifications
Broadband Pent up Demand

- *"The most pent up demand for wireless technology is for Video, Digital Imaging, Wireless Report Management System, Mapping / Geographic Information System, and Automatic Vehicle Location."*
- Survey demand of broadband applications
  - **94%** requested digital imaging
  - **97%** requested mapping/GIS
  - **91%** requested streaming video

Application	Question	Yes	No	Total	Yes %	No %
Automatic Vehicle Location	Use	16	21	37	43.24%	56.76%
	Use Wirelessly	8	24	32	25.00%	75.00%
	Want Wirelessly	28	5	33	84.85%	15.15%
Computer Aided Dispatch	Use	27	10	37	72.97%	27.03%
	Use Wirelessly	3	21	24	12.50%	87.50%
	Want Wirelessly	12	10	22	54.55%	45.45%
Digital Imaging	Use	29	8	37	78.38%	21.62%
	Use Wirelessly	3	32	35	8.57%	91.43%
	Want Wirelessly	31	2	33	93.94%	6.06%
E-mail	Use	33	4	37	89.19%	10.81%
	Use Wirelessly	20	15	35	57.14%	42.86%
	Want Wirelessly	33	4	37	89.19%	10.81%
Mapping / Geographic Info. System	Use	28	8	36	77.78%	22.22%
	Use Wirelessly	8	19	27	29.63%	70.37%
	Want Wirelessly	30	1	31	96.77%	3.23%
Remote Database Access / Data Entry	Use	22	15	37	59.46%	40.54%
	Use Wirelessly	12	22	34	35.29%	64.71%
	Want Wirelessly	25	6	31	80.65%	19.35%
Report Management System	Use	20	11	31	64.52%	35.48%
	Use Wirelessly	0	26	26	0.00%	100.00%
	Want Wirelessly	17	8	25	68.00%	32.00%
Text Messaging	Use	32	6	38	84.21%	15.79%
	Use Wirelessly	26	11	37	70.27%	29.73%
	Want Wirelessly	32	4	36	88.89%	11.11%
Video	Use	17	20	37	45.95%	54.05%
	Use Wirelessly	4	31	35	11.43%	88.57%
	Want Wirelessly	31	3	34	91.18%	8.82%

Table 4: Technology Deployment and Demand

RWBN Operations Cost Comparison Budget

Yearly Operational Cost for 700 MHz Public Safety Networks

		Total/Year
3	*Node B Spares/Year*	$ 514,800.00
4	*Core Network Spares/Year*	$ 210,000.00
6	*Generators & UPS Spares/Year*	$ 147,600.00
8	*Antennas Spares/Year*	$ 12,818.00
9	*Site Lease/Year*	$ 1,104,000.00
10	*Site Utilities/Year*	$ 748,800.00
12	*Microwave Spares*	$ 32,500.00
13	*Devices Depreciation/Year (35,000 Subs)*	$ 2,187,500.00
14	*Devices Spare/Year*	$ 437,500.00
15	*Customer Operations Personnel -*	$ 400,000.00
16	*Help Desk $5/Sub*	$ 175,000.00
17	*Network Monitoring Engineers*	$ 600,000.00
18	*Trouble Shooting Crews - 1 Person per County*	$ 800,000.00
19	*Software Upgrade*	$ 300,000.00
20	*Test Lab Engineer*	$ 100,000.00
21	*Test Lab Equipment Upgrade*	$ 25,000.00
20	*Vendor 24/7 Tier 3 Support*	$ 100,000.00
21	*Training*	$ 150,000.00
22	*Other Contract Support*	$ 100,000.00
23	*Roaming Fees*	$ 315,000.00
	Total	$ 8,460,518.00

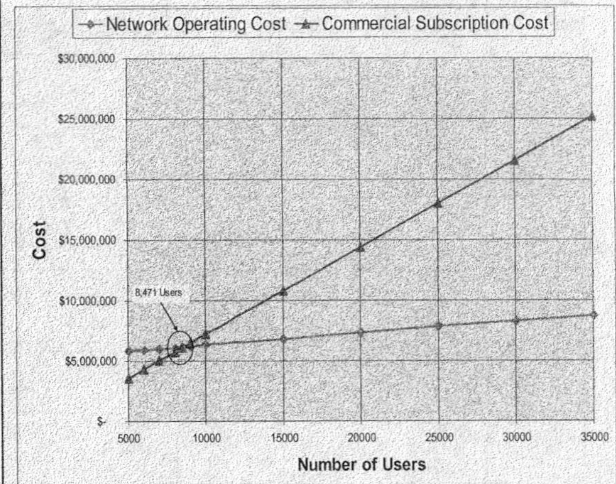

- Assumes annual NCR operation of 104 sites and 35,000 subscriber devices for $8.5 million.
- Cost comparison with commercial services assumes $60/mo per device equally $25 million annually for 35,000 subscribers.
- Breakeven subscribers equal 8,471.

What if we don't do RWBN today?

- If a major incident would to occur today, the vast majority of first responder's mobile data computing will fail.
- We will pay $16M more a year to commercial carriers than it will cost to operate the network ourselves. (assumes 35,000 first responders in NCR)

Or

- We will cube ourselves!

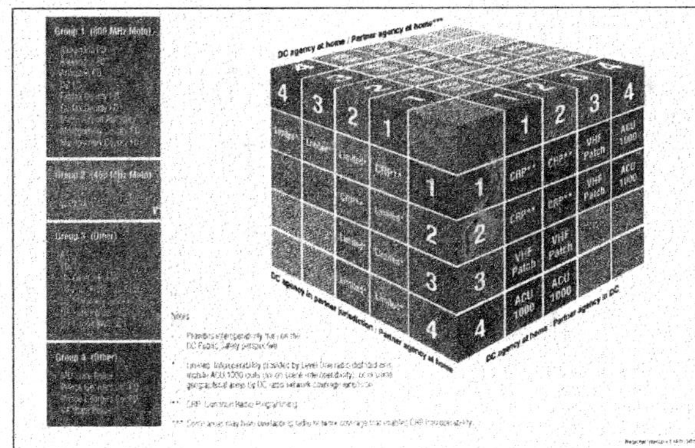

ONE HUNDRED NINTH CONGRESS

Congress of the United States

House of Representatives

COMMITTEE ON GOVERNMENT REFORM

2157 RAYBURN HOUSE OFFICE BUILDING
WASHINGTON, DC 20515–6143

MAJORITY (202) 225–5074
FACSIMILE (202) 225–3974
MINORITY (202) 225–5051
TTY (202) 225–6852

http://reform.house.gov

June 27, 2006

Mr. Edward D. Reiskin
Deputy Mayor for Public Safety and Justice
John A. Wilson Building
1350 Pennsylvania Avenue, NW
Suite 327
Washington, DC 20004

Dear Mr. Reiskin:

Thank you for your testimony before the Committee on Government Reform on June 15, 2006. The Committee greatly appreciated your participation in the discussion about the Urban Area Security Initiative grant process and looks forward to a continued dialogue with the representatives of the National Capital Region about steps being taken to make the capital area safe. This letter sets forth additional questions for the record for the June 15 hearing. Please answer the following questions:

1. The NCR's effectiveness rating was in the bottom 25 percent of all Urban Areas. Reportedly, the proposals were viewed as less innovative and less likely to produce sustained, high-impact results than those submitted by other areas.

   Prior to receiving feedback from DHS on the NCR application (which is scheduled for this month), and based upon further review of the application, what do you see as possible weaknesses associated with the investment justifications?

   What specific changes would you make in any re-submittals of investment justifications?

2. The NCR jurisdictions are currently drafting a NCR strategic plan. Thus, a strategic plan was not available to guide the NCR application for funding.

   To what extent did the lack of a strategic plan impair NCR's preparation of its funding application? Please explain.

How will the completed strategic plan provide the needed justification for future federal funding applications?

Additionally, how is this plan different from the Enhancement Plan? The Enhancement Plan lists vague needs such as, "Additional equipment is needed to protect citizens from attack," and "People need to train regularly." How is the strategic plan more specific, and when will it be released? Likewise, did the grant application to DHS clarify generalities such as these? Please provide examples.

3. Mr. Robertson's testimony stated the NCR has "developed emergency transportation plans."

   Do these plans include a plan to evacuate METRO passengers in the event of an emergency (something the NCR Enhancement Plan indicated has not yet been addressed)? If not, why? If so, please include a brief summary of the plans in place.

4. Mr. Robertson's testimony also indicates the Senior Policy Group and the Chief Administrative Officers will now prioritize needs and develop a list of projects that can be funded at the current award level. When do you anticipate this list being finished?

   Are there contingency plans for the NCR jurisdictions, acting as a body or independently, to address priorities that will not receive federal funding from UASI? If not, why? If so, please include a brief summary of these priorities.

5. In response to a GAO report released in May 2004, the National Capital Region Senior Policy Group stated, "The Deputy Mayor for Public Safety and Justice (DMPSJ), Executive Office of the Mayor of the Government of the District of Columbia has been tasked by the Mayor to lead the District's efforts relating to Homeland Security. In this regard, the Deputy Mayor is responsible for oversight of all spending related to special appropriations, regional planning for Homeland Security, inter- and intra- agency planning (with federal, state, and local governments), and the development of a Homeland Security Strategy for the District."

   When you were appointed to this position in January 2005, did you assume control of the areas outlined above? If not, who does assume this control?

   If so, can you please provide the Committee with an itemized list of how "special appropriations" for Homeland Security have been spent on behalf of the District of Columbia since you took office, particularly the UASI grants and the Department of Defense Emergency Supplemental Appropriations?

6. The Senior Policy Group's response to GAO also indicated the DMPSJ is the "Homeland Security point of contact through which all grant monies must be

administered. Additionally, the District was chosen as the Administrative Agent for grants awarded to the NCR."

Are you also responsible for overseeing and tracking grants, such as UASI grants and the DOD Emergency Supplemental Appropriations, for the NCR? If not, who is responsible for this oversight and tracking?

If so, can you please provide the Committee with an itemized list of how all federal grants for Homeland Security (including UASI grants and the DOD Emergency Supplemental Appropriations), from January 2005 to the present, have been spent on behalf of the National Capital Region? Please be sure to include the amount distributed by the District, as the Administrative Agent, to each NCR jurisdiction and for what purpose, as well as a list of actual expenditures by each jurisdiction that received grant monies.

How was this spending of federal grants coordinated across all the NCR jurisdictions, and will any changes in this process be made with respect to grants currently being issued to the NCR?

Please respond to this letter no later than 5:00 p.m. on July 7, 2006. Your response and any supporting documents should be delivered to Room 2157 Rayburn House Office Building. If you have any questions about this letter, please contact Wimberly Fair at 202-225-5074.

Thank you again for your time and assistance with these requests.

Sincerely,

Tom Davis
Chairman
Committee on Government Reform

ONE HUNDRED NINTH CONGRESS

Congress of the United States

House of Representatives

COMMITTEE ON GOVERNMENT REFORM

2157 RAYBURN HOUSE OFFICE BUILDING
WASHINGTON, DC 20515–6143

MAJORITY (202) 225–5074
FACSIMILE (202) 225–3974
MINORITY (202) 225–5051
TTY (202) 225–6852

http://reform.house.gov

June 27, 2006

Mr. Robert Crouch
Assistant to the Governor for Commonwealth Preparedness
Commonwealth of Virginia
1111 E. Broad Street, 3rd Floor
Richmond, VA 23219

Dear Mr. Crouch:

Thank you for your testimony before the Committee on Government Reform on June 15, 2006. The Committee greatly appreciated your participation in the discussion about the Urban Area Security Initiative grant process and looks forward to a continued dialogue with the representatives of the National Capital Region about steps being taken to make the capital area safe. This letter sets forth additional questions for the record for the June 15 hearing. Please answer the following questions:

1. The NCR's effectiveness rating was in the bottom 25 percent of all Urban Areas. Reportedly, the proposals were viewed as less innovative and less likely to produce sustained, high-impact results than those submitted by other areas.

   Prior to receiving feedback from DHS on the NCR application (which is scheduled for this month), and based upon further review of the application, what do you see as possible weaknesses associated with the investment justifications?

   What specific changes would you make in any re-submittals of investment justifications?

2. The NCR jurisdictions are currently drafting a NCR strategic plan. Thus, a strategic plan was not available to guide the NCR application for funding.

   To what extent did the lack of a strategic plan impair NCR's preparation of its funding application? Please explain.

How will the completed strategic plan provide the needed justification for future federal funding applications?

Additionally, how is this plan different from the Enhancement Plan? The Enhancement Plan lists vague needs such as, "Additional equipment is needed to protect citizens from attack," and "People need to train regularly." How is the strategic plan more specific, and when will it be released? Likewise, did the grant application to DHS clarify generalities such as these? Please provide examples.

3. Mr. Robertson's testimony stated the NCR has "developed emergency transportation plans."

   Do these plans include a plan to evacuate METRO passengers in the event of an emergency (something the NCR Enhancement Plan indicated has not yet been addressed)? If not, why? If so, please include a brief summary of the plans in place.

4. Mr. Robertson's testimony also indicates the Senior Policy Group and the Chief Administrative Officers will now prioritize needs and develop a list of projects that can be funded at the current award level. When do you anticipate this list being finished?

   Are there contingency plans for the NCR jurisdictions, acting as a body or independently, to address priorities that will not receive federal funding from UASI? If not, why? If so, please include a brief summary of these priorities.

Please respond to this letter no later than 5:00 p.m. on July 7, 2006. Your response and any supporting documents should be delivered to Room 2157 Rayburn House Office Building. If you have any questions about this letter, please contact Wimberly Fair at 202-225-5074.

Thank you again for your time and assistance with these requests.

Sincerely,

Tom Davis

Tom Davis
Chairman
Committee on Government Reform

ONE HUNDRED NINTH CONGRESS

Congress of the United States

House of Representatives

COMMITTEE ON GOVERNMENT REFORM

2157 RAYBURN HOUSE OFFICE BUILDING

WASHINGTON, DC 20515–6143

MAJORITY (202) 225–5074
FACSIMILE (202) 225–3974
MINORITY (202) 225–5051
TTY (202) 225–6852

http://reform.house.gov

June 27, 2006

Mr. Dennis Schrader
Director
Office of Homeland Security
State of Maryland
State House
Annapolis, MD 21401

Dear Mr. Schrader:

Thank you for your testimony before the Committee on Government Reform on June 15, 2006. The Committee greatly appreciated your participation in the discussion about the Urban Area Security Initiative grant process and looks forward to a continued dialogue with the representatives of the National Capital Region about steps being taken to make the capital area safe. This letter sets forth additional questions for the record for the June 15 hearing. Please answer the following questions:

1. The NCR's effectiveness rating was in the bottom 25 percent of all Urban Areas. Reportedly, the proposals were viewed as less innovative and less likely to produce sustained, high-impact results than those submitted by other areas.

   Prior to receiving feedback from DHS on the NCR application (which is scheduled for this month), and based upon further review of the application, what do you see as possible weaknesses associated with the investment justifications?

   What specific changes would you make in any re-submittals of investment justifications?

2. The NCR jurisdictions are currently drafting a NCR strategic plan. Thus, a strategic plan was not available to guide the NCR application for funding.

   To what extent did the lack of a strategic plan impair NCR's preparation of its funding application? Please explain.

How will the completed strategic plan provide the needed justification for future federal funding applications?

Additionally, how is this plan different from the Enhancement Plan? The Enhancement Plan lists vague needs such as, "Additional equipment is needed to protect citizens from attack," and "People need to train regularly." How is the strategic plan more specific, and when will it be released? Likewise, did the grant application to DHS clarify generalities such as these? Please provide examples.

3. Mr. Robertson's testimony stated the NCR has "developed emergency transportation plans."

   Do these plans include a plan to evacuate METRO passengers in the event of an emergency (something the NCR Enhancement Plan indicated has not yet been addressed)? If not, why? If so, please include a brief summary of the plans in place.

4. Mr. Robertson's testimony also indicates the Senior Policy Group and the Chief Administrative Officers will now prioritize needs and develop a list of projects that can be funded at the current award level. When do you anticipate this list being finished?

   Are there contingency plans for the NCR jurisdictions, acting as a body or independently, to address priorities that will not receive federal funding from UASI? If not, why? If so, please include a brief summary of these priorities.

Please respond to this letter no later than 5:00 p.m. on July 7, 2006. Your response and any supporting documents should be delivered to Room 2157 Rayburn House Office Building. If you have any questions about this letter, please contact Wimberly Fair at 202-225-5074.

Thank you again for your time and assistance with these requests.

Sincerely,

Tom Davis

Tom Davis
Chairman
Committee on Government Reform

ONE HUNDRED NINTH CONGRESS

Congress of the United States

House of Representatives

COMMITTEE ON GOVERNMENT REFORM
2157 RAYBURN HOUSE OFFICE BUILDING
WASHINGTON, DC 20515–6143

MAJORITY (202) 225–5074
FACSIMILE (202) 225–3974
MINORITY (202) 225–5051
TTY (202) 225–6852

http://reform.house.gov

June 27, 2006

Mr. Edward D. Reiskin
Deputy Mayor for Public Safety and Justice
John A. Wilson Building
1350 Pennsylvania Avenue, NW
Suite 327
Washington, DC 20004

Dear Mr. Reiskin:

Thank you for your testimony before the Committee on Government Reform on June 15, 2006. The Committee greatly appreciated your participation in the discussion about the Urban Area Security Initiative grant process and looks forward to a continued dialogue with the representatives of the National Capital Region about steps being taken to make the capital area safe. This letter sets forth additional questions for the record for the June 15 hearing. Please answer the following questions:

1. The NCR's effectiveness rating was in the bottom 25 percent of all Urban Areas. Reportedly, the proposals were viewed as less innovative and less likely to produce sustained, high-impact results than those submitted by other areas.

   Prior to receiving feedback from DHS on the NCR application (which is scheduled for this month), and based upon further review of the application, what do you see as possible weaknesses associated with the investment justifications?

   What specific changes would you make in any re-submittals of investment justifications?

2. The NCR jurisdictions are currently drafting a NCR strategic plan. Thus, a strategic plan was not available to guide the NCR application for funding.

   To what extent did the lack of a strategic plan impair NCR's preparation of its funding application? Please explain.

How will the completed strategic plan provide the needed justification for future federal funding applications?

Additionally, how is this plan different from the Enhancement Plan? The Enhancement Plan lists vague needs such as, "Additional equipment is needed to protect citizens from attack," and "People need to train regularly." How is the strategic plan more specific, and when will it be released? Likewise, did the grant application to DHS clarify generalities such as these? Please provide examples.

3. Mr. Robertson's testimony stated the NCR has "developed emergency transportation plans."

   Do these plans include a plan to evacuate METRO passengers in the event of an emergency (something the NCR Enhancement Plan indicated has not yet been addressed)? If not, why? If so, please include a brief summary of the plans in place.

4. Mr. Robertson's testimony also indicates the Senior Policy Group and the Chief Administrative Officers will now prioritize needs and develop a list of projects that can be funded at the current award level. When do you anticipate this list being finished?

   Are there contingency plans for the NCR jurisdictions, acting as a body or independently, to address priorities that will not receive federal funding from UASI? If not, why? If so, please include a brief summary of these priorities.

Please respond to this letter no later than 5:00 p.m. on July 7, 2006. Your response and any supporting documents should be delivered to Room 2157 Rayburn House Office Building. If you have any questions about this letter, please contact Wimberly Fair at 202-225-5074.

Thank you again for your time and assistance with these requests.

Sincerely,

Tom Davis

Tom Davis
Chairman
Committee on Government Reform

ONE HUNDRED NINTH CONGRESS

Congress of the United States

House of Representatives

COMMITTEE ON GOVERNMENT REFORM

2157 RAYBURN HOUSE OFFICE BUILDING

WASHINGTON, DC 20515–6143

MAJORITY (202) 225–5074
FACSIMILE (202) 225–3974
MINORITY (202) 225–5051
TTY (202) 225–6852

http://reform.house.gov

June 21, 2006

The Honorable George Foresman
Undersecretary for Preparedness
U.S. Department of Homeland Security
Washington, D.C. 20528

Dear Undersecretary Foresman:

I want to thank you for your testimony before the Committee on Government Reform's hearing entitled, "Regional Insecurity: DHS Grants to the National Capital Area" on June 15, 2006. Your presentation was informative and provided members of the Committee valuable information.

At the hearing you committed to respond in writing to several questions posed by members of the Committee. This letter sets forth those questions and request for information for the record. We kindly ask for a response to the following questions:

1) The District's application was not rated highly. What was deficient in its application compared to other jurisdictions?

2) Did the Department take into account the daytime population in the District of Columbia when calculating State grants?

3) Did the peer reviewers have security clearance and access to real-time threat analysis?

Lastly, the Committee requested a copy of the contract between the Department and Booz Allen concerning its participation in facilitating the peer review process.

Please respond to this letter no later than 5:00 p.m. on July 7, 2006. Your response and any supporting documents should be delivered to Room 2157, Rayburn House Office Building. If you have any questions about this letter, please contact Keith Ausbrook at 202-225-5074.

Thank you for your participation in the hearing and your assistance with these requests.

Sincerely,

Tom Davis
Chairman

Office of the Under Secretary for Preparedness
U.S. Department of Homeland Security
Washington, DC 20528

Homeland Security

JUL 2 8 2006

The Honorable Tom Davis
Chairman
Committee on Government Reform
U.S. House of Representatives
Washington, D.C. 20515

Dear Mr. Chairman:

Thank you for the opportunity to appear before the House Committee on Government Reform and to discuss the Department of Homeland Security's (Department) Fiscal Year (FY) 2006 Homeland Security Grant Program (HSGP). I was especially pleased to discuss how HSGP, and its component program, the Urban Areas Security Initiative (UASI), affects the National Capital Region (NCR) and the District of Columbia.

The overall mission of the HSGP is to work towards greater preparedness for the Nation's homeland security. Thus, the program is designed to align available resources with national priorities and the National Preparedness Goal. Its primary goal is to enhance the capabilities of State and local governments to prevent, protect, respond to, and recover from a disaster or emergency. As a result, we worked closely with stakeholders to provide guidance about the program and communicate the process and standards/criteria used in making allocations.

As you are aware, the funds appropriated to the overall HSGP were significantly less in FY 2006. In fact, there was $600 million less for the overall HSGP and about $125 million less for the specific UASI program. This meant the amount available for allocation among each State, territory and urban area was reduced.

Nevertheless, it is important to note that the Department has provided more than $213 million to the National Capital Region since 2003. This figure translates to about eight percent of the UASI funds provided, which is consistent with the historical average. Further, the Department will provide the District of Columbia an additional $7.4 million in resources through the FY 2006 State Homeland Security Program (SHSP) and Law Enforcement Terrorism Prevention Program (LETPP). Similarly, the Department will provide the Commonwealth of Virginia an additional $15 million and the State of Maryland $14 million in SHSP and LETPP resources to help prepare, protect, and respond to incidents within the NCR.

Following the hearing you submitted three questions for additional information, which are addressed below.

Question No. 1:
The District's application was not rated highly. What was deficient compared to other jurisdictions?

Response:
Through the grant allocation process, the Department made every effort to ensure that the allocation determinations were driven by measurable facts and impartial risk-based analyses. We sought the highest level of integrity and transparency in the grant process, so that all requesters and stakeholders knew the standards for evaluation beforehand, to support development of their investment justifications. These steps were designed to ensure both strengthened national homeland security along with fiscal responsibility.

Of the $1.7 billion allotted to the HSGP programs, approximately $1.3 billion was allocated based upon two primary factors: risk and effectiveness. The remaining $400 million were pre-determined base allocations mandated by Congressional direction.

Risk was the primary factor accounting for nearly 70 percent of the funding allocations. In past years, the risk element was largely determined by both population size and density. Over time, we have been able to utilize more sophisticated techniques to analyze risk consistent with the evolution of national homeland security efforts. In FY 2006, the risk factor was comprised of three primary components: Threat, Vulnerability, and Consequence. The Threat component represents an adversary's intent to attack a specific target and its potential capability to execute the attack; the Vulnerability component embodies the susceptibility to an adversary's attack and the likelihood that it will achieve an impact; and the Consequence component measures the possible impact from such an attack. Further, we analyzed the risk factors both on the basis of the specific characteristics of risk to a particular asset, as well as risk inherent to a geographical area.

Based upon these various factors and considerations, we collected and compiled literally millions of data points and billions of calculations derived from various sources within the law enforcement, first-responder, private-sector, and intelligence communities. As a result of all of this information and the resulting calculations, each jurisdiction was assigned a risk factor based upon the best information available and in accordance with Congressional guidance. We will, of course, continue to work to enhance the data used in the risk analysis process.

The second factor, effectiveness, evaluates the use of the grant funds consistent with Congressional intent. Our responsibility is to ensure that the limited funds were allocated in the most effective way possible. It is important to keep in mind that HSGP funding is considered an investment to manage risk throughout the Nation. Funding resources target solutions, which will produce tangible results, raises the overall baseline for national preparedness, and achieve the best return on the Nation's investment of limited resources.

In order to meet this objective, we asked that applicants address effectiveness concerns in their submissions, essentially making a "business case" for grant funding. This business case, or investment justification, allowed us to evaluate how urban areas and States were tying funding to address pre-identified needs and comply with program requirements.

A panel of experts helped evaluate proposals from requestors. This Peer Review process established objectivity and consistency, and ensured its integrity and credibility. The panel of approximately 100 reviewers represented nearly every geographic corner of the country, including 48 States and territories, 38 urban areas, and 11 reviewers from various Federal agencies. In addition, the reviewers represented a wide range of expertise from law enforcement personnel, firefighters, first-responders, emergency managers, and State Homeland Security

Advisors: seasoned professionals from programmatic, policy, and operational disciplines. These Peer Reviewers are experts in the field, since they experience the State and local homeland security mission on the ground every single day.

The Department provided the Peer Reviewers with the same guidance and standards previously provided to the requesting stakeholders prior to their submissions. The Peer Reviewers then assessed the various proposals and assigned a numerical value for each submission. After receiving these ratings, the Department ran statistical analyses to determine if there were any signs of bias from the Peer Reviewers or the process itself. Indeed, the results yielded a normal distribution indicating no signs of bias. Indeed, in the survey of Peer Reviewers, nearly 83 percent felt that the process produced objective, consistent, and reliable results.

A combination of the risk analyses and the effectiveness ratings from the Peer Reviewers were assessed, in order to achieve a final determination on grant funding allocations. We were able to assess the risks that each jurisdiction faced in conjunction with the best use of available resources in a fair and objective manner and allocate grant funding accordingly.

This combination of the risk analysis and effectiveness ratings from the Peer Reviewers drove the allocations decisions for the District of Columbia and all other FY 2006 HSGP recipients.

Overall, based on the peer review panel evaluation, the District's application for funding through the State grant programs was rated in the bottom 50 percent of all State submissions. The submission from the National Capital Region for funding through the UASI program placed within the bottom 25 percent of all urban area submissions.

To add more detail to this response, I have enclosed three documents which the Committee may find useful. The first two are "For Official Use Only" documents titled "FY 2006 Homeland Security Grant Program (HSGP) Award for the District of Columbia" and "FY 2006 Urban Areas Security Initiative (UASI) Award for the National Capital Region." These documents, which were individually tailored for and provided to each of the FY 20006 HSGP recipients, provide a more detailed assessment of the effectiveness analysis of the investment proposals from the District of Columbia and the National Capital Region. They also provide more detail as to the Department's geographic and asset risk analysis. The third document, titled "Allocation Methodology," provides a discussion of how the overall risk ranking and effectiveness ranking came together to drive the FY 2006 allocation decisions. It is my hope that the Committee finds these materials useful.

Question No. 2:
Did the Department take into account the daytime population in the District of Columbia when calculating State grants?

Response:
As described previously, the risk analysis was based on three principal variables: threat, vulnerability, and consequence. The model included both asset-based and geographically-based terrorist risk calculations. That is, the analysis considered both the risks to specific types of assets as well as the risks inherent to a particular geographic area. In evaluating geographically-based risk, DHS evaluated reported threats, law enforcement activity (using Federal Bureau of Investigation and Immigration and Customs Enforcement terrorism case data), and suspicious incidents reported during the evaluation period. Vulnerability factors for each geographic area,

such as the area's proximity to international borders, were also evaluated. Lastly, DHS estimated the potential consequences of an attack on that area, including human health, economy, strategic mission, and psychological impacts. In analyzing the consequences on human health, DHS considered population and population density, including transient populations such as commuters and visitors in urban areas. Thus, transient populations were included in the risk analysis and in the overall FY 2006 HSGP allocation process for the District of Columbia.

Question No. 3:
Did the peer reviewers have security clearance and access to real-time threat analysis?

Response:
The goal of the peer review process was to determine the anticipated effectiveness of an applicant's proposed investments when considered against a set of uniform and objective criteria derived from the National Preparedness Goal and the Targeted Capabilities List. The role of the peer reviewer was to determine effectiveness scores for each applicant's overall submission and to determine whether the various proposals represented a sound investment of federal dollars. To fulfill their role, the peer reviewers did not require either security clearances or access to real-time threat analysis as they were not involved in the risk analysis position of the allocation process, which accounted for almost 70 percent of an applicant's final allocation. To assist in the discussion of the role of the peer reviewers, I have enclosed a document titled "Peer Review Process" which the Committee may find useful.

With regard to your request for a copy of the contract between the Department and Booz Allen Hamilton (BAH), we would note that the Department did not have a direct contractual relationship with BAH for this requirement. Instead, the Department entered into an Inter-Agency Agreement with the Department of Defense (DoD), through which the Department was able to leverage a pre-existing DoD contract for these services. Therefore, we have referred your request for the contract and the task order to DoD. Nevertheless, we have included a copy of the Inter-Agency Agreement between the Department and DoD. As this document incorporates information that may be exempt from public release under the Freedom of Information Act, we request your assistance in protecting it accordingly.

Once again Mr. Chairman, the Department has made every effort to ensure that our grants program is driven by measurable facts and impartial, risk-based assessments and the Department is committed to continuously improving these processes and the administration of these vital programs. This year's analysis was by far the most comprehensive since the Department was created. We thank you for your commitment to this issue and look forward to continuing to work with you and your colleagues on the Committee on Government Reform to better secure our Nation.

Sincerely,

George W. Foresman
Under Secretary

Mr. Chairman - please let me know if you feel more information is needed.

Enclosures (5)

FY 2006 Homeland Security Grant program (HSGP) Award for the District of Columbia
FY 2006 Homeland Security Grant program (HSGP) Award for the NCR
Allocation Methodology
Peer Review Process
Inter-Agency Agreement for Booz Allen Hamilton services

U.S. Department of Homeland Security
Washington, DC 20528

Homeland Security

FY 2006 Homeland Security Grant Program (HSGP) Award for the District of Columbia

District of Columbia's FY 2006 HSGP Award Information:

State Homeland Security Program (SHSP)	$ 4,270,000
Urban Areas Security Initiative (UASI)	$ 46,470,000
Law Enforcement Terrorism Prevention Program (LETPP)	$ 3,110,000
Metropolitan Medical Response System (MMRS)	n/a
Citizen Corps Program (CCP)	$ 165,142
TOTAL	**$ 54,015,142**

Risk: The District of Columbia was among the bottom 25% of all States and Territories based on the DHS comparative risk analysis

Effectiveness: The District of Columbia's FY 2006 proposed solutions were rated among the bottom 50% of all States and Territories by the peer review panels

DHS Grant Funding History for the District of Columbia

Since 9/11, the District of Columbia has received the following grant funding from DHS[1]:

FY 2005	$ 114,237,851
FY 2004	$ 52,623,684
FY 2003	$ 85,963,541
FY 2002	$ 6,454,625
FY 2002 – FY 2005 TOTAL	**$ 259,279,701**

Comparing the District of Columbia FY 2005 and FY 2006 Funding

The appropriated funding for the overall HSGP decreased by approximately 29 percent from FY 2005 to FY 2006, as illustrated in Figure 1. The table below provides a comparison on HSGP award information for the District of Columbia in FY 2005 and FY 2006. The table includes the amounts as a percentage of the total available HSGP funds. In addition, the table also provides an illustrative example of what the District of Columbia's 2005 HSGP funding would have been using the same allocation methodology used last year with the total funds available in FY 2006. While this figure is to be used only for comparison purposes, it provides a useful basis for comparing awards received in different fiscal years with different totally HSGP program funding amounts.

Figure 1: Total HSGP Funding

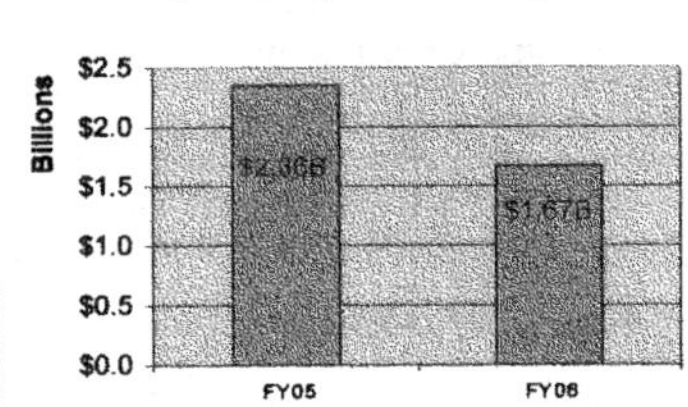

	FY 2005	FY 2006	Difference From FY 2005 to FY 2006
District of Columbia HSGP Award Amount[2]	$ 94,640,300	$ 54,015,142	($ 40,625,158)
Comparable FY 2005 HSGP Award in FY 2006 Dollars	$ 76,233,856		($ 22,218,714)
HSGP Award as Percentage of Total Funding Available	4.04%	3.23%	(0.81%)

[1] Totals reflect funding from the Office of Grants and Training, Federal Emergency Management Agency, and the Transportation Sec urity Administration.

[2] For comparison purposes, funding from the FY 2005 Emergency Manage ment Performance Grant program has not been included in the FY 2005 total, as it was awarded as a separate program in FY 2006.

Summary of the FY 2006 Homeland Security Grant Program (HSGP) Risk Analysis for the District of Columbia

Overview

As part of this risk-based approach to preparedness, DHS' Fiscal Year 2006 (FY06) risk methodology represents a major step forward in the analysis of the risk of terrorism, resulting in the most accurate estimation to-date of the relative risk faced by our Nation's communities. In response to State and local partner feedback, the FY06 methodology incorporates a number of significant enhancements over previous years' analyses, including:

- Incorporation of strategic threat analysis from the Intelligence Community
- Improved attribution of threat and law enforcement activity data
- Greater depth and breadth in critical infrastructure and key asset data
- Inclusion of populated areas outside official city limits to encourage regionalization
- Incorporation of transient populations, such as tourists and commuters

DHS defines risk by three principal variables: ***threat***, or the likelihood of a type of attack that might be attempted, ***vulnerability***, or the likelihood that an attacker would succeed with a particular attack type, and ***consequence***, or the potential impact of a particular attack.

The risk model used to allocate HSGP funds includes both asset-based and geographically-based terrorist risk calculations. DHS combines these complementary risk calculations to produce an estimate of the relative risk of terrorism faced by a given area.

Asset-based risk – The asset-based approach uses strategic threat estimates from the Intelligence Community of an adversary's intent and capability to attack different types of assets (such as chemical plants, stadiums, and commercial airports) using different attack methods. DHS analyzes the vulnerability of each asset type relative to each attack method to determine the form of attack most likely to be successful.

Additionally, DHS estimates the consequences that successful attacks would have on each asset type, including human health, economic, strategic mission, and psychological impacts. This analysis yields a relative risk estimate for each asset type, which DHS applies to a given geographic area, based on the number of each asset type present within that area.

Geographically-based risk – The geographic-based approach allows DHS to consider general characteristics of a geographic area mostly independent of the assets that exist within that area. First, DHS evaluates reported threats, law enforcement activity (using Federal Bureau of Investigation and Immigration and Customs Enforcement terrorism case data), and suspicious incidents reported during the evaluation period. Next, DHS considers vulnerability factors for each geographic area, such as the area's proximity to international borders.

Lastly, DHS estimates the potential consequences of an attack on that area, including human health (e.g., population, population density, transient populations), economy (e.g., percentage of Gross Domestic Product, total agriculture sales, international cargo value), strategic mission (e.g., defense industrial base), and psychological impacts.

District of Columbia Results

- In evaluating the relative risk, DHS looked at data for the District of Columbia. DHS then evaluated both the risk to individual assets residing within the District, and the risk to the geographic area itself, the sum of which placed the District into the top 25%, top 50%, bottom 50%, or bottom 25% of all States and Territories considered in the DHS comparative risk analysis.

- A total of 513 assets in ten critical infrastructure sectors met the criteria for consideration in the analysis; for the District of Columbia, the largest concentration of these was Commercial Assets. When the asset-based risk analysis was applied to these assets, the District of Columbia fell in the bottom 25% of the 56 States and Territories, meaning that the risk associated with individual assets in the District of Columbia was lower than three-quarters of the States and Territories.

- In the geographic-based risk analysis, the District of Columbia fell in the top 50% of the 56 States and Territories, indicating that it had a higher level of risk associated with reported threat and investigative activity, resident and visitor populations, and other geographic criteria than at least half of the States and Territories.

- When the asset-based and geographic-based risk data for the District of Columbia was aggregated, it placed it in the bottom 25% of the 56 States and Territories.

For Official Use Only (FOUO)

DISTRICT OF COLUMBIA

STATE TOTAL RISK = Bottom 25%

Considered together, the sum of the numeric asset-based and geographically-based risk values provides a calculation of the total risk to the State, with the top 25% corresponding to those States that were considered to be at a higher level of risk than at least three-quarters of all States considered in the DHS comparative risk analysis. The bottom 25% corresponds to those States that were considered to be at a lower level of risk than at least three-quarters of all States considered in the DHS comparative risk analysis.

STATE ASSET RISK = Bottom 25%

Asset risk is the numerical value that is the product of consequence, vulnerabiilty, and threat associated with a specific asset. Asset risk is calculated for each unique asset, as well as being summed for all assets associated with a specific candidate in a given grant to determine candidate asset risk. The Asset-based Risk percentage indicates the relative risk to assets based on the results of the risk analysis.

ASSET INFORMATION	COUNT
BANKING AND FINANCE	0
CHEMICAL AND HAZARDOUS MATERIALS INDUSTRY	20
COMMERCIAL ASSETS	314
DAMS	0
ENERGY	3
HEALTHCARE AND PUBLIC HEALTH	16
NATIONAL MONUMENTS AND ICONS	18
NUCLEAR FACILITIES	4
POSTAL AND SHIPPING	45
TELECOMMUNICATIONS	30
TRANSPORTATION	61
WATER	2
Total Assets	513

STATE GEOGRAPHIC RISK = Top 50%

Geographic risk is the numerical value of the product of consequence, vulnerability, and threat associated with a specific candidate. Consequence, vulnerability, and threat scores rely on inherent attributes of the geographic candidate. The Geographic-based Risk percentage indicates the relative risk to populations and geographic areas for the State based on the results of the risk analysis.

GEOGRAPHIC ATTRIBUTES		COUNT
Human Consequences	Population Density - State	9,296
	Population-State 2004 Est.	553,523
Economic Consequences	Gross Domestic Product (GDP)	75,264,000,000
	State International Export Trade	1,164,327,394
	State Total Agriculture Sales	0
Mission Consequences	Defense Industrial Base Facilities	Classified
	Military Bases	12
	National Health Stockpile Sites	0
Psychological Consequences	Special Events	6
Subtotal Geographic Risk	Sum of UASI Population in State	553,523
	UASI Pop Density in State	1,467
Vulnerability	International Border	0
	Length of Nuclear WIPP route within state	0
	Port of Entry/Border Crossings Total Throughput	1,155,139
	Ratio of Population to Law Enforcement	116
Threat	FBI Cases	Classified
	I-94 Visitors Countries of Interest Destination City	Classified
	Intelligence Community Reports	Classified
	Immigration and Customs Enforcement Cases	Classified
	Port of Entry/Border Crossings I-94 Countries of Interest	Classified
	Suspicious Incidents	Classified

ASSET KEY

There were 40 State Asset Types considered for the asset-based risk calculations. The asset types were grouped into twelve sectors which each have specific thresholds and criteria for inclusion in the calculations.

Sector	Asset Types
BANKING AND FINANCE	Financial Facilities (>$8 billion)
CHEMICAL AND HAZARDOUS MATERIALS INDUSTRY	Chemical Manufacturing Facilities (DHS Tier 1, 2, 3)
COMMERCIAL ASSETS	Colleges and Universities, Convention Centers (>250,000 sq. ft.), Enclosed Shopping Malls (>100 stores), Hotel Casinos (>500 rooms & >1,000 empl./shift), Primary And Secondary Schools, Stadiums (Non University >35,000 seating capacity), Tall Commercial Buildings (>600 ft.), and Theme Parks (>1,000,000 visitors/yr.)
DAMS	Dams (USACOE "high hazard") and Levees
ENERGY	Electricity Generation Facilities (>500MV or 1,500 MW), Electricity Substations (>230 KV), LNG Terminals, Natural Gas Compressor Stations (>20 in.), Petroleum Pumping Stations (>20 in.), Petroleum Refineries, and Petroleum Storage Tank Farms (>1,000,000 barrels)
HEALTHCARE AND PUBLIC HEALTH	Hospitals and National Health Stockpile Sites
NATIONAL MONUMENTS AND ICONS	National Monuments and Icons
NUCLEAR FACILITIES	Non Power Nuclear Reactors, Nuclear Power Plants, and Nuclear Research Labs
POSTAL AND SHIPPING	Commercial Overnight Shipping Facilities
TELECOMMUNICATIONS	Telcomm-Telephone Hotels (>5 exchanges) and Trans Oceanic Cable Landings (>250 MHz)
TRANSPORTATION	Commercial Airports, Ferry Terminals - Buildings, Maritime Port Facilities, Mass Transit Commuter Rail and Subway Stations, Railroad Bridges, Railroad Passenger Stations, Railroad Tunnels, Road Commuter Tunnels, and Road Bridges (>100,000 vehicles/day)
WATER	Potable Water Treatment Facilities >3,300 customers)

For Official Use Only (FOUO)

Summary of the FY 2006 Homeland Security Grant Program (HSGP) Effectiveness Analysis for the District of Columbia

Overview

For the FY 2006 HSGP, DHS employed a peer review process to evaluate the anticipated effectiveness of the proposed solutions in grant applications to address identified needs. Over 100 reviewers representing 48 States and Territories, 38 Urban Areas, and two Federal agencies participated in the review process. The outcome of this process was a set of objective, consistent, and defensible effectiveness scores for the overall submission and for each individual Investment.

The overall submission was evaluated on the following criteria:

- ***Relevance to the Interim National Preparedness Goal Implementation*** – The extent to which the overall submission shows alignment with National Priorities, Target Capabilities, Program and Capability Enhancement Plan Initiatives, and goals and objectives from the State/Urban Area Homeland Security Strategies, as well as the extent to which the proposed Investments as a whole will help the Nation to be more prepared.
- ***Connection to the Enhancement Plan*** – The extent to which the overall submission relates back to the Program and Capability Enhancement Plan to show that the applicant is committed to addressing its priority needs.
- ***Complete Picture*** – The extent to which the individual Investments relate to each other to portray a complete picture of plans for the homeland security program.
- ***Innovativeness*** – The extent to which solutions presented in the Investment were the result of thoughtful planning, consideration, and creativity.
- ***Feasibility and Reasonableness*** – The extent to which solutions presented in the overall submission can be implemented and are appropriately scoped given the planned level of effort, and the extent to which the budget request aligns with the size and scope of the proposed Investments.

The individual Investments were evaluated on the following criteria:

- ***Relevance*** – The relationship of the Investment to the tenets of the Interim National Preparedness Goal. Relevance is gauged through the Investment's connection to the National Priorities, Target Capabilities List, State/Urban Area Homeland Security Strategy goals and objectives, and Initiatives from the Program and Capability Enhancement Plan.
- ***Regionalization*** – The ability to communicate, plan, and collaborate across disciplines and jurisdictions to leverage scarce resources for common solutions. Regionalization encourages States and Urban Areas to coordinate preparedness activities more effectively within and across jurisdictional boundaries by spreading costs, pooling resources, sharing risk, and increasing the value of their preparedness Investments through collaborative efforts.
- ***Impact*** - The effect that implementing (or not implementing) an Investment has on risk. Impact addresses the effect that the Investment will have on addressing threats, vulnerabilities, and/or consequences of catastrophic events that applicants might face.

- ***Sustainability*** – The ability to sustain a target capability once the benefits of an Investment are achieved through identification of funding sources that can be used beyond the current grant period.
- ***Implementation Approach*** – Demonstration that the appropriate people, resources, and tools are (or will be) in place to manage the Investment, address priorities, and deliver results though appropriate expenditure of any funding received.

District of Columbia Results

The peer review panel evaluated the District of Columbia's overall submission and individual Investments based on the aforementioned criteria. For every Investment submitted, the panel members reviewed and scored the responses provided for each question. These scores were then averaged to compute a score for each individual Investment. The individual Investment scores were averaged to produce an average Investment score. The peer review panel also determined the overall submission score based on their review of the entire application. The average Investment score was combined with the overall submission score to determine the final effectiveness score.

Below are the summary results based on the peer review panel's evaluation of the District of Columbia's submission. Several types of information are provided: the District's performance relative to all other submissions based on the total effectiveness score and specific results of peer reviewer evaluation of the District's submission. With the exception of the total effectiveness score, the ratings included in the summary sheet do not reflect results relative to other applicants. They are based on the peer reviewers' independent evaluation of the District's submission based upon the prescribed criteria.

In June 2006, DHS will provide feedback and recommendations from the peer review panels on this submission.

- The District of Columbia's submission for the FY 2006 HSGP is in the bottom 50% of all State submissions.

- As presented in the Investment Summary table on the following page, the District of Columbia submitted a total of 9 Investments. The peer review panel considered *Critical Infrastructure Protection Program* to be the strongest Investment, and considered *Mass Care* to be the Investment most in need of improvement.

- As presented in the Overall Submission table, the peer review panel considered the District of Columbia's overall submission to be above average in the areas of Relevance to Goal Implementation, Complete Picture, and Feasibility and Reasonableness, and average in Connection to the Enhancement Plan, and Innovativeness.

- The Investment Detail table provides a breakdown of Investment performance by category to aid the District of Columbia in identifying areas of strength as well as areas needing improvement.

Investment Justification Special Conditions

The bottom 15% of all State Investments, as scored through the peer review process, will have a special condition governing the drawdown of funds associated with these Investments included in the grant award. The District of Columbia did not have any Investments in the bottom 15%, and, as such, this special condition is not applicable to the District of Columbia.

 For Official Use Only (FOUO)

DISTRICT OF COLUMBIA

TOTAL EFFECTIVENESS = Bottom 50%

Investment Summary

This table ranks the Investments that were included in the FY 2006 HSGP submission according to evaluation results, from those that were strongest to those most needing improvement. Additional details regarding Investment evaluation results are provided below in the Investment Detail table.

Investment Name	Rank
Critical Infrastructure Protection Program	1
Information Sharing	2
Incident Response	3
Citizen Preparedness	4
Planning	5
Law Enforcement Investigations and Operations	6
Communications	7
Medical Surge and Mass Prophylaxis	8
Mass Care	9

Overall Submission

This table summarizes the evaluation results of the FY 2006 HSGP submission as a whole. Five categories were used in this overall evaluation.

I. Relevance to Goal Implementation	II. Connection to the Enhancement Plan	III. Complete Picture	IV. Innovativeness	V. Feasibility and Reasonableness
Above Average	Average	Above Average	Average	Above Average

Legend

Needs Improvement | Below Average | Average | Above Average | Excellent

Investment Detail

This table details individual Investment evaluation results, broken down into five categories, to aid applicants in identifying areas of strength and areas needing improvement.

Investment Name	I. Relevance	II. Regionalization	III. Impact	IV. Sustainability	V. Implementation Approach
Critical Infrastructure Protection Program	Above Average	Above Average	Excellent	Average	Above Average
Information Sharing	Above Average	Above Average	Excellent	Average	Above Average
Incident Response	Above Average	Above Average	Above Average	Average	Above Average
Citizen Preparedness	Above Average	Above Average	Excellent	Average	Above Average
Planning	Above Average	Above Average	Above Average	Average	Above Average
Law Enforcement Investigations and Operations	Above Average	Above Average	Above Average	Below Average	Above Average
Communications	Above Average	Above Average	Above Average	Average	Average
Medical Surge and Mass Prophylaxis	Above Average	Above Average	Above Average	Average	Average
Mass Care	Above Average	Above Average	Above Average	Below Average	Above Average

 For Official Use Only (FOUO)

FOR OFFICIAL USE ONLY

U.S. Department of Homeland Security
Washington, DC 20528

Homeland Security

FY 2006 Urban Areas Security Initiative (UASI) Award for the National Capital Region

National Capital Region Area FY 2006 UASI Award **$ 46,470,000**

Risk: The National Capital Region Area was among the top 25% of all Urban Areas based on the FY 2006 DHS comparative risk analysis

Effectiveness: The National Capital Region Area's FY 2006 proposed solutions were rated among the bottom 25% of all Urban Areas by the peer review panels

UASI Funding History for the National Capital Region Area

Since 9/11, the National Capital Region Area has received the following funding through the UASI program:

FY 2005	$ 77,500,000
FY 2004	$ 29,128,623
FY 2003	$ 60,490,851
FY 2003 – FY 2005 TOTAL	**$ 167,119,474**

Comparing the National Capital Region Area's FY 2005 and FY 2006 UASI Awards

The appropriated funding for the overall UASI program decreased by more than 14 percent from FY 2005 to FY 2006, as illustrated in Figure 1. The table below provides a comparison of UASI award information for the National Capital Region Area in FY 2005 and FY 2006. The table includes the amounts awarded to the National Capital Region Area through the program in FY 2005 and FY 2006, and the award amounts as a percentage of the total available UASI funds. In addition, the table also provides an illustrative example of what the National Capital Region Area's FY 2005 UASI funding would have been using the same allocation methodology used last year with the total funds available in FY 2006. While this figure is to be used only for comparison purposes, it provides a useful basis for comparing awards received in different fiscal years with different total UASI program funding amounts.

Figure 1: Total UASI Funding

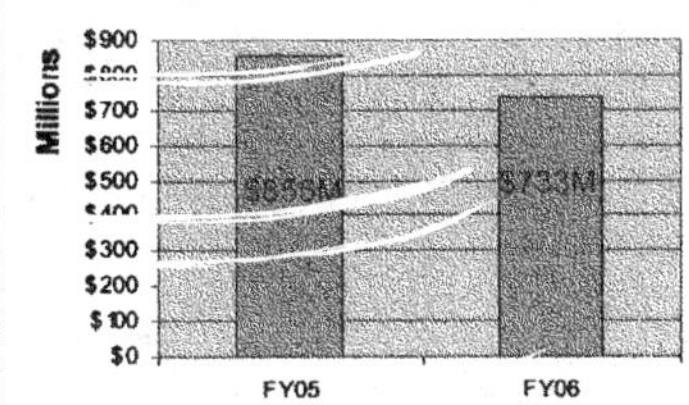

	FY 2005	FY 2006	Difference From FY 2005 to FY 2006
National Capital Region Area UASI Award Amount	$ 77,500,000	$ 46,470,000	($ 31,030,000)
Comparable FY 2005 UASI Award in FY 2006 Dollars	$ 66,380,711		($ 19,910,711)
UASI Award as Percentage of Total UASI Funding Available	9.34%	6.54%	(2.80%)

Summary of the FY 2006 Homeland Security Grant Program (HSGP) Risk Analysis for the National Capital Region

Overview

As part of this risk-based approach to preparedness, DHS' Fiscal Year 2006 (FY06) risk methodology represents a major step forward in the analysis of the risk of terrorism, resulting in the most accurate estimation to-date of the relative risk faced by our Nation's communities. In response to State and local partner feedback, the FY06 methodology incorporates a number of significant enhancements over previous years' analyses, including:

- Incorporation of strategic threat analysis from the Intelligence Community
- Improved attribution of threat and law enforcement activity data
- Greater depth and breadth in critical infrastructure and key asset data
- Inclusion of populated areas outside official city limits to encourage regionalization
- Incorporation of transient populations, such as tourists and commuters

DHS defines risk by three principal variables: ***threat***, or the likelihood of a type of attack that might be attempted, ***vulnerability***, or the likelihood that an attacker would succeed with a particular attack type, and ***consequence***, or the potential impact of a particular attack.

The risk model used to allocate HSGP funds includes both asset-based and geographically-based terrorist risk calculations. DHS combines these complementary risk calculations to produce an estimate of the relative risk of terrorism faced by a given area.

Asset-based risk – The asset-based approach uses strategic threat estimates from the Intelligence Community of an adversary's intent and capability to attack different types of assets (such as chemical plants, stadiums, and commercial airports) using different attack methods. DHS analyzes the vulnerability of each asset type relative to each attack method to determine the form of attack most likely to be successful.

Additionally, DHS estimates the consequences that successful attacks would have on each asset type, including human health, economic, strategic mission, and psychological impacts. This analysis yields a relative risk estimate for each asset type, which DHS applies to a given geographic area, based on the number of each asset type present within that area.

Geographically-based risk – The geographic-based approach allows DHS to consider general characteristics of a geographic area mostly independent of the assets that exist within that area. First, DHS evaluates reported threats, law enforcement activity (using Federal Bureau of Investigation and Immigration and Customs Enforcement terrorism case data), and suspicious incidents reported during the evaluation period. Next, DHS considers vulnerability factors for each geographic area, such as the area's proximity to international borders.

Lastly, DHS estimates the potential consequences of an attack on that area, including human health (e.g., population, population density, transient populations), economy (e.g., percentage of Gross Domestic Product, total agriculture sales, international cargo value), strategic mission (e.g., defense industrial base), and psychological impacts.

National Capital Region Results

- In evaluating the relative risk to the National Capital Region, DHS looked at data for the National Capital Region and a ten mile buffer extending from the limits of that area. DHS then evaluated both the risk to individual assets residing within that combined area, and the risk to the geographic area itself, the sum of which placed the Urban Area into the top 25%, top 50%, bottom 50%, or bottom 25% of the 46 eligible Urban Areas.

- A total of 3,417 assets in eleven critical infrastructure sectors met the criteria for consideration in the analysis; for the National Capital Region, the largest concentration of these was Commercial Assets. When the asset-based risk analysis was applied to these assets, the National Capital Region fell in the top 25% of the eligible Urban Areas, meaning that the risk associated with individual assets in the National Capital Region was higher than at least three-quarters of the eligible Urban Areas.

- In the geographic-based risk analysis, the National Capital Region fell in the top 25% of the eligible Urban Areas, indicating that it had a higher level of risk associated with reported threat and investigative activity, resident and visitor populations, and other geographic criteria than at least three-quarters of the eligible Urban Areas.

- When the asset-based and geographic-based risk data for the National Capital Region was aggregated, it placed it in the top 25% of all eligible Urban Areas.

- Based on the DHS comparative risk analysis, the National Capital Region placed in the top 35 areas.

NATIONAL CAPITAL REGION

URBAN AREA TOTAL RISK = Top 25%

Considered together, the sum of the numeric asset-based and geographically-based risk values provides a calculation of the total risk to the Urban Area, with the top 25% corresponding to those Urban Areas that were considered to be at a higher level of risk than at least three-quarters of the 46 eligible Urban Areas considered in the DHS comparative risk analysis. The bottom 25% corresponds to those Urban Areas that were considered to be at a lower level of risk than at least three-quarters of the 46 eligible Urban Areas considered in the DHS comparative risk analysis.

URBAN AREA ASSET RISK = Top 25%

Asset risk is the numerical value that is the product of consequence, vulnerability, and threat associated with a specific asset. Asset risk is calculated for each unique asset, as well as being summed for all assets associated with a specific candidate in a given grant to determine candidate asset risk. The Asset-based Risk percentage indicates the relative risk to assets for the Urban Area based on the results of the risk analysis.

ASSET INFORMATION	COUNT
BANKING AND FINANCE	2
CHEMICAL AND HAZARDOUS MATERIALS INDUSTRY	38
COMMERCIAL ASSETS	2,154
DAMS	0
ENERGY	93
HEALTHCARE AND PUBLIC HEALTH	128
NATIONAL MONUMENTS AND ICONS	18
NUCLEAR FACILITIES	7
POSTAL AND SHIPPING	511
TELECOMMUNICATIONS	150
TRANSPORTATION	293
WATER	23
Total Assets	3,417

URBAN AREA GEOGRAPHIC RISK = Top 25%

Geographic risk is the numerical value of the product of consequence, vulnerability, and threat associated with a specific candidate. Consequence, vulnerability, and threat scores rely on inherent attributes of the geographic candidate. The Geographic-based Risk percentage indicates the relative risk to populations and geographic areas for the Urban Area based on the results of the risk analysis.

GEOGRAPHIC ATTRIBUTES		COUNT
Human Consequences	Population City - Commuter	4,529,837
	Population-City 10 Mile Buffer Zone Night Census	4,639,574
	Population-Visitor	133,377
	Urban Area	3,075.74
Mission Consequences	Defense Industrial Base Facilities	Classified
	Military Bases	43
Psychological Consequences	Special Events	0
Inherited Geographic Risk	Population-City Limit Census	4,511,032
	Sum of Port Population in City	0
	Average Daily Rail Ridership	672,682
Vulnerability	Port of Entry/Border Crossings Total Throughput	913,266
	Length of Nuclear WIPP route within city buffer	3.23
Threat	FBI Cases	Classified
	I-94 Visitors Countries of Interest Destination City	Classified
	Intelligence Community Reports	Classified
	Immigration and Customs Enforcement Cases	Classified
	Port of Entry/Border Crossings I-94 Countries of Interest	Classified
	Suspicious Incidents	Classified
	Vessels of Interest	Classified

ASSET KEY

There were 38 Urban Area Asset Types considered for the asset-based risk calculations. The asset types were grouped into twelve sectors which each have specific thresholds and criteria for inclusion in the calculations.

Sector	Asset Types
BANKING AND FINANCE	Financial Facilities (>$8 billion)
CHEMICAL AND HAZARDOUS MATERIALS INDUSTRY	Chemical Manufacturing Facilities (DHS Tier 1, 2, 3)
COMMERCIAL ASSETS	Colleges and Universities, Convention Centers (>250,000 sq. ft.), Enclosed Shopping Malls (>100 stores), Hotel Casinos (>500 rooms & >1,000 empl./shift), Primary And Secondary Schools, Stadiums (Non University >35,000 seating capacity), Tal Commercial Buildings (>600 ft.), and Theme Parks (>1,000,000 visitors/yr.)
DAMS	Dams (USACOE "high hazard") and Levees
ENERGY	Electricity Generation Facilities (>500MV or 1,500 MW), Electricity Substations (>230 KV), LNG Terminals, Natural Gas Compressor Stations (>20 in.), Petroleum Pumping Stations (>20 in.), Petroleum Refineries. and Petroleum Storage Tank Farms (>1,000,000 barrels)
HEALTHCARE AND PUBLIC HEALTH	Hospitals and National Health Stockpile Sites
NATIONAL MONUMENTS AND ICONS	National Monuments and Icons
NUCLEAR FACILITIES	Non Power Nuclear Reactors, Nuclear Power Plants, and Nuclear Research Labs
POSTAL AND SHIPPING	Commercial Overnight Shipping Facilities
TELECOMMUNICATIONS	Telcomm-Telephone Hotels (>5 exchanges) and Trans Oceanic Cable Landings (>250 MHz)
TRANSPORTATION	Commercial Airports, Ferry Terminals - Buildings, Maritime Port Facilities, Mass Transit Commuter Rail and Subway Stations, Railroad Bridges, Railroad Passenger Stations, Railroad Tunnels, Road Commuter Tunnels, and Road Bridges (>100,000 vehicles/day)
WATER	Potable Water Treatment Facilities >3,300 customers)

Summary of the FY 2006 Homeland Security Grant Program (HSGP) Effectiveness Analysis for the National Capital Region

Overview

For the FY 2006 HSGP, DHS employed a peer review process to evaluate the anticipated effectiveness of the proposed solutions in grant applications to address identified needs. Over 100 reviewers representing 48 States and Territories, 38 Urban Areas, and two Federal agencies participated in the review process. The outcome of this process was a set of objective, consistent, and defensible effectiveness scores for the overall submission and for each individual Investment.

The overall submission was evaluated on the following criteria:

- ***Relevance to Interim National Preparedness Goal Implementation*** – The extent to which the overall submission shows alignment with National Priorities, Target Capabilities, Program and Capability Enhancement Plan Initiatives, and goals and objectives from the State/Urban Area Homeland Security Strategies, as well as the extent to which the proposed Investments as a whole will help the Nation to be more prepared.
- ***Connection to the Enhancement Plan*** – The extent to which the overall submission relates back to the Program and Capability Enhancement Plan to show that the applicant is committed to addressing its priority needs.
- ***Complete Picture*** – The extent to which the individual Investments relate to each other to portray a complete picture of plans for the homeland security program.
- ***Innovativeness*** – The extent to which solutions presented in the Investment were the result of thoughtful planning, consideration, and creativity.
- ***Feasibility and Reasonableness*** – The extent to which solutions presented in the overall submission can be implemented and are appropriately scoped given the planned level of effort, and the extent to which the budget request aligns with the size and scope of the proposed Investments.

The individual Investments were evaluated on the following criteria:

- ***Relevance*** – The relationship of the Investment to the tenets of the Interim National Preparedness Goal. Relevance is gauged through the Investment's connection to the National Priorities, Target Capabilities List, State/Urban Area Homeland Security Strategy goals and objectives, and Initiatives from the Program and Capability Enhancement Plan.
- ***Regionalization*** – The ability to communicate, plan, and collaborate across disciplines and jurisdictions to leverage scarce resources for common solutions. Regionalization encourages States and Urban Areas to coordinate preparedness activities more effectively within and across jurisdictional boundaries by spreading costs, pooling resources, sharing risk, and increasing the value of their preparedness Investments through collaborative efforts.
- ***Impact*** - The effect that implementing (or not implementing) an Investment has on risk. Impact addresses the effect that the Investment will have on addressing threats, vulnerabilities, and/or consequences of catastrophic events that applicants might face.

- ***Sustainability*** – The ability to sustain a target capability once the benefits of an Investment are achieved through identification of funding sources that can be used beyond the current grant period.
- ***Implementation Approach*** – Demonstration that the appropriate people, resources, and tools are (or will be) in place to manage the Investment, address priorities, and deliver results though appropriate expenditure of any funding received.

National Capital Region Results

The peer review panel evaluated the National Capital Region's overall submission and individual Investments based on the aforementioned criteria. For every Investment submitted, the panel members reviewed and scored the responses provided for each question. These scores were then averaged to compute a score for each individual Investment. The individual Investment scores were averaged to produce an average Investment score. The peer review panel also determined the overall submission score based on their review of the entire application. The average Investment score was combined with the overall submission score to determine the final effectiveness score.

Below are the summary results based on the peer review panel's evaluation of the National Capital Region's submission. Several types of information are provided: the Urban Area's performance relative to all other submissions based on the total effectiveness score and specific results of peer reviewer evaluation of the Urban Area's submission. With the exception of the total effectiveness score, the ratings included in the summary sheet do not reflect results relative to other applicants. They are based on the peer reviewers' independent evaluation of the Urban Area's submission based upon the prescribed criteria.

In June 2006, DHS will provide feedback and recommendations from the peer review panels on this submission.

- The National Capital Region's submission for the FY 2006 HSGP is in the bottom 25% of all Urban Area submissions.

- As presented in the Investment Summary table on the following page, the National Capital Region submitted a total of 12 Investments. The peer review panel considered *NCR Regional Collaboration for Citizen Protection & Preparedness* to be the strongest Investment, and considered *NCR Regional Incident Planning & Response* to be the Investment most in need of improvement.

- As presented in the Overall Submission table, the peer review panel considered the National Capital Region's overall submission to be above average in Relevance to Goal Implementation, and below average in Innovativeness. The peer review panel considered the National Capital Region's overall submission to be average in Connection to the Enhancement Plan, Complete Picture, and Feasibility and Reasonableness.

- The Investment Detail table provides a breakdown of Investment performance by category to aid the National Capital Region in identifying areas of strength as well as areas needing improvement.

Investment Justification Special Conditions

The bottom 15% of all Urban Area Investments, as scored through the peer review process, will have a special condition governing the drawdown of funds associated with these Investments included in the grant award. The following Investment proposed by the National Capital Region scored among the bottom 15% of all 478 Urban Area Investments as determined through the effectiveness peer review process.

- *NCR Regional Incident Planning & Response*

A special condition will be included in the grant award prohibiting the drawdown of funds against this Investment until it has been approved through DHS. Information in the effectiveness summary sheet as well as feedback from the peer review panels, to be provided under separate cover, should be utilized as a tool to update the Investment.

NATIONAL CAPITAL REGION

URBAN AREA TOTAL EFFECTIVENESS = Bottom 25%

Investment Summary

This table ranks the investments that were included in the FY 2006 HSGP submission according to evaluation results, from those that were strongest to those most needing improvement. Additional details regarding Investment evaluation results are provided below in the Investment Detail table.

Investment Name	Rank
NCR Regional Collaboration for Citizen Protection & Preparedness	1
NCR Regional Medical Response and Support	2
NCR Regional Interoperable Communications	3
NCR Regional Mass Care	4
NCR Regional Law Enforcement Collaboration	5
NCR Regional CBRNE Threat Detection	6
NCR Regional Explosive Device Operations	7
NCR Critical Infrastructure Protection	8
NCR Regional Information Sharing	9
NCR Regional Logistics & Distribution	10
NCR WMD/Haz-Mat Collaboration	11
NCR Regional Incident Planning & Response	12

Overall Submission

This table summarizes the evaluation results of the FY 2006 HSGP submission as a whole. Five categories were used in this overall evaluation.

I. Relevance to Goal Implementation	II. Connection to the Enhancement Plan	III. Complete Picture	IV. Innovativeness	V. Feasibility and Reasonableness
Above Average	Average	Average	Below Average	Average

Legend

Needs Improvement	Below Average	Average	Above Average	Excellent

Investment Detail

This table details individual Investment evaluation results, broken down into five categories, to aid applicants in identifying areas of strength and areas needing improvement.

Investment Name	I. Relevance	II. Regionalization	III. Impact	IV. Sustainability	V. Implementation Approach
NCR Regional Collaboration for Citizen Protection & Preparedness	Above Average	Above Average	Above Average	Average	Above Average
NCR Regional Medical Response and Support	Above Average	Above Average	Above Average	Average	Above Average
NCR Regional Interoperable Communications	Above Average	Above Average	Above Average	Average	Average
NCR Regional Mass Care	Above Average	Above Average	Above Average	Average	Above Average
NCR Regional Law Enforcement Collaboration	Above Average	Above Average	Above Average	Average	Average
NCR Regional CBRNE Threat Detection	Above Average	Above Average	Above Average	Average	Average
NCR Regional Explosive Device Operations	Above Average	Above Average	Above Average	Average	Average
NCR Critical Infrastructure Protection	Above Average	Above Average	Above Average	Average	Average
NCR Regional Information Sharing	Above Average	Above Average	Above Average	Average	Average
NCR Regional Logistics & Distribution	Above Average	Above Average	Above Average	Average	Average
NCR WMD/Haz-Mat Collaboration	Above Average	Above Average	Above Average	Average	Average
NCR Regional Incident Planning & Response	Average	Above Average	Above Average	Average	Average

Homeland Security Grant Program

Preparedness Directorate
Office of Grants and Training
Tracy A. Henke, Assistant Secretary

Allocation Methodology

Maximizing the Impact of Homeland Security Funding

In Fiscal Year 2006 (FY06), the U.S. Department of Homeland Security (DHS) introduced a new allocation methodology for evaluating applications under the Homeland Security Grant Program (HSGP). The new methodology represents DHS' most comprehensive approach to date for prioritizing homeland security resources. For the first time, DHS is able to align HSGP resources with the National Priorities and target capabilities established by the Interim National Preparedness Goal.

*In Fiscal Year 2006, DHS' Homeland Security Grant Program (HSGP) will award approximately **$1.7 billion** to States, Territories, and Urban Areas to enhance their ability to prevent, protect against, respond to, and recover from terrorist attacks, major disasters, and other emergencies.*

The methodology bases HSGP allocations primarily on two factors:

1. an analysis of relative ***risk*** to assets as well as risk to populations and geographic areas
2. the anticipated ***effectiveness*** of State and Urban Area grant proposals in addressing their identified homeland security needs.

These factors are used to determine allocation amounts for the State Homeland Security Program (SHSP), the Urban Areas Security Initiative (UASI), and the Law Enforcement Terrorism Prevention Program (LETPP) under HSGP. In addition, a base award is allocated under SHSP and LETPP according to the USA PATRIOT Act formula. Risk and effectiveness also are used to allocate all UASI funding.

More Comprehensive Risk Analysis Used in FY06

The DHS FY06 risk methodology represents a major step forward in the analysis of the risk of terrorism, based on the most comprehensive analysis to date of the relative risk faced by our Nation's communities. DHS defines risk by three principal variables: ***threat***, or the likelihood of a type of attack that might be attempted, ***vulnerability***, or the likelihood that an attacker would succeed with a particular attack type, and ***consequence***, or the potential impact of a particular attack.

Asset-Based Risk
+ Geographic-Based Risk
Total Risk for State/Urban Area

The FY06 risk model included two distinct risk calculations: risk to assets and risk to populations and geographic areas. DHS combined these complementary risk calculations to produce an estimate of the relative risk of terrorism faced by a given area.

Peer Reviewers Determined Anticipated Effectiveness Scores

States and Urban Areas used the Investment Justification to formally request FY06 HSGP funding for SHSP, LETPP, and UASI. In FY06, more than 100 peer reviewers read the Investment Justifications and worked independently to determine a preliminary effectiveness score before convening in panels to discuss the findings of their review, develop final scores, and provide comments on each submission. The reviewers evaluated submissions based on specific criteria, including relevance, regionalization, sustainability, implementation approach, and impact. Each submission was reviewed and scored in two different ways, resulting in an average score for the Individual Investments and an overall score for the submission. DHS combined the average score of the individual Investments with the overall submission score, as determined by the peer review panel, to determine the final effectiveness score.

This approach to evaluating anticipated effectiveness seeks to recognize applicants for proposing relevant, innovative, and reasonable investments that will directly impact our Nation's preparedness.

Average Individual Investment Score
+ Overall Investment Justification Score
Effectiveness Score for State/Urban Area

Integrating Risk and Effectiveness Scores to Maximize Impact on Preparedness

In order to allocate funding based upon the relative risk and anticipated effectiveness scores, DHS plotted each applicant on a two-by-two matrix. Based on the combination of their risk and effectiveness scores, applicants fell into one of four categories on the matrix:

1. higher risk, higher effectiveness
2. higher risk, lower effectiveness
3. lower risk, higher effectiveness
4. lower-risk, lower effectiveness

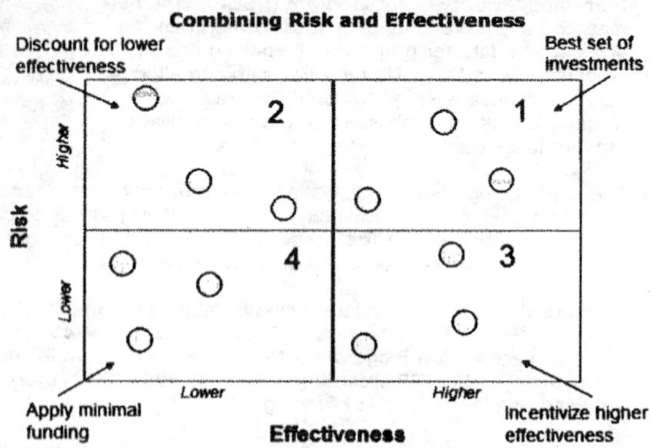

In allocating funds to each grouping of candidates, DHS targeted resources towards those areas of our Nation at greatest risk, while still substantively rewarding the significant efforts undertaken by applicants in presenting effective solutions. Each applicant's final funding allocation was determined using a combination of its risk and effectiveness scores, with a two-thirds weight applied to risk and one-third weight applied to effectiveness. This weighting further reinforces DHS' commitment to a risk-based approach to national preparedness.

The approximately $1.7 billion in HSGP funds allocated to States, Territories, and Urban Areas this year is intended to help strengthen capabilities, enhance preparedness planning, and ultimately reduce the impact of major events on lives, property and the economy. The methodology DHS employed to allocate FY06 HSGP funds was purposefully designed to help us reduce our national risk by targeting funding at high risk areas, while at the same time encouraging and recognizing those solutions that will effectively build a preparedness baseline across the Nation.

Homeland Security Grant Program

Preparedness Directorate
Office of Grants and Training
Tracy A. Henke, Assistant Secretary

Peer Review Process

Involving State and Local Professionals in Scoring Grant Applications

*In Fiscal Year 2006, DHS' Homeland Security Grant Program (HSGP) will award approximately **$1.7 billion** to States, Territories, and Urban Areas to enhance their ability to prevent, protect against, respond to, and recover from terrorist attacks, major disasters, and other emergencies.*

The U.S. Department of Homeland Security (DHS) engages State and local partners in evaluating and scoring all Fiscal Year 2006 (FY06) Homeland Security Grant Program (HSGP) applications. Stakeholder involvement helps to enhance the thoroughness of this process and makes the program more reflective of local needs.

The peer review scoring process was conducted to assess how effective the solutions proposed in the HSGP applications address the applicants' homeland security program needs. States and Urban Areas applying for HSGP funds must not only identify and prioritize their homeland security needs, but also demonstrate effectiveness of proposed solutions to reduce risk. Proposed solutions are described in detail in the applicant's Investment Justification. Each State and Urban Area was allowed to submit up to 15 Investments in their Investment Justifications.

In FY06, more than 100 peer reviewers, including incident managers, fire chiefs, law enforcement personnel, grant managers, and representatives from State and local emergency management agencies, participated in the peer review process. The peer review process took place at the National Emergency Training Center in Emmitsburg, Maryland during the week of April 2, 2006. The resulting scores indicate the effectiveness of proposed solutions to addressing identified homeland security needs.

Ensuring Credibility in the Selection Process

The peer review selection process is designed to ensure a fair and equitable peer review. The DHS Office of Grants and Training, which oversees the HSGP, solicited potential peer review participants from each State and Territory, including the various Urban Areas, along with information on the nominees' background and experience. DHS then developed a preliminary list of the most qualified candidates based on the following criteria:

- Extensive subject-matter expertise across multiple homeland security disciplines
- Experience managing an integrated homeland security program or initiative
- Experience in the policy, operations, and programmatic arenas
- Familiarity with the HSGP

DHS submitted the names of potential reviewers to State Administrative Agencies responsible for homeland security in each State, to make a final determination of peer reviewers for the FY06 HSGP grant cycle. *State homeland security professionals made the final determination for FY06 HSGP Peer Review Panelists.*

Building Diversity into Panel Composition

To ensure that peer review panels were made up of reviewers from varied backgrounds and experience—and to avoid potential conflicts of interest—diversity was emphasized in forming the panels. Each panel was comprised of up to seven reviewers, which included up to three reviewers representing States and Territories, up to three representing Urban Areas, and up to one reviewer representing federal agencies.

The composition of each panel was balanced with representation from eastern, western, and central States, as well as States with large and small populations.

Investment Justifications from Urban Areas were grouped with corresponding State applications to enable reviewers to evaluate the coordination between both efforts. Reviewers were not, however, assigned to review Investment Justifications from their own States or (to the extent possible) those of neighboring States or Territories.

Scoring Process

Initially, peer reviewers worked independently to determine a preliminary score of applicants' proposed solutions to address identified needs. In the second phase, reviewers convened in panels to discuss the findings of their review, develop final scores, and comment on each Investment Justification. Each panel was responsible for scoring six Investment Justifications. Each reviewer was responsible for submitting a final score as consensus among panel members was not necessary.

The Investment Justifications were scored in two ways. First, an Individual Investment Score was based on 17 questions, each with detailed criteria, addressing issues such as:

- Relevance of the proposed solution
- Regionalization
- Anticipated impact
- Sustainability of the proposed Investment
- Implementation approach

Second, the Overall Investment Justification Score was based on these criteria:

- Relevance to the Interim National Preparedness Goal
- Connection to the spirit and scope of the Program and Capability Enhancement Plan
- Extent to which individual Investments related to each other
- Innovativeness in approaching identified needs
- Feasibility and reasonableness of the proposed Investments

The combination of the Individual Investment Score and Overall Investment Justification Score constitute the final Effectiveness score.

For additional information please contact the Centralized Scheduling and Information Desk at 1-800-368-6498 or askcsid@dhs.gov.

INTERAGENCY AGREEMENT

1. IAA NO.: HSHQDC-05-X-00021

PAGE 1 OF 2

2. ORDER NO.:

3. REQUISITION NO.: RODP-05-01059

4. SOLICITATION NO.:

5. EFFECTIVE DATE: 06/02/2005

6. AWARD DATE: 06/02/2005

7. PERIOD OF PERFORMANCE: 06/15/2005 TO 06/14/2006

8. SERVICING AGENCY
Dept. of Defense, Defense Technical Information Center
Information Analysis Center
8725 John J. Kingman Road
Suite 0944
Attn: [redacted]
Ft. Belvoir VA 22060-6218
POC
TELEPHONE NO.

9. DELIVER TO
Office for Domestic Preparedness
810 Seventh Street NW
Washington DC 20531
USA

10. REQUESTING AGENCY
Department of Homeland Security-OPO
245 Murray Drive
Bldg. 410
Office of Procurement Operations
Attn: [redacted]
Washington DC 20528
POC
TELEPHONE NO. 202 357-8341

11. INVOICE OFFICE
U. S. Department of Justice
Office of Justice Programs
Accounting Division
810 Seventh Street, NW
Washington DC 20531

12. ISSUING OFFICE
Department of Homeland Security
Office of Procurement Operations
245 Murray Drive
Bldg. 410
Washington DC 20528

13. LEGISLATIVE AUTHORITY: Economy Act, 31 U.S.C. 1535 & 1536

14. PROJECT ID:

15. PROJECT TITLE:

16. ACCOUNTING DATA: See Schedule

17. ITEM NO.	18. SUPPLIES/SERVICES	19. QUANTITY	20. UNIT	21. UNIT PRICE	22. AMOUNT
	Delivery: 06/14/2006				
0001	The purpose of this agreement is to fund technical area tasks that provide studies, analysis, data gathering and other unique research and development activities in support of DHS Office of State and Local Government Coordination and Preparedness programs. Contracting Fee: 4% Accounting Info: 5 T 12 25 00 00 Funded: $4,377,985.31 Continued ...				4,637,985.31

23. PAYMENT PROVISIONS

24. TOTAL AMOUNT: $4,637,985.31

25a. SIGNATURE OF GOVERNMENT REPRESENTATIVE (SERVICING)

25b. NAME AND TITLE

25c. DATE

26a. SIGNATURE OF GOVERNMENT REPRESENTATIVE (REQUESTING)

26b. CONTRACTING OFFICER [redacted]

26c. DATE 06/13/05

IAA NO	ORDER NO	PAGE	OF
HSHQDC-05-X-00021		2	2

Accounting Info:
5 T TH 25 00 00
Funded: $260,000.00

Department of Defense (DOD), Defense Technical and Information Center (DTIC) under the Survivability/Vulnerability Information Analysis Center (SURVIAC)

Contract No. SP0700-03-D-1380

Total amount of award: $7,100,000.00.